CAMBRIDGE LIBRARY COLLECTION

Books of enduring scholarly value

Religion

For centuries, scripture and theology were the focus of prodigious amounts of scholarship and publishing, dominated in the English-speaking world by the work of Protestant Christians. Enlightenment philosophy and science, anthropology, ethnology and the colonial experience all brought new perspectives, lively debates and heated controversies to the study of religion and its role in the world, many of which continue to this day. This series explores the editing and interpretation of religious texts, the history of religious ideas and institutions, and not least the encounter between religion and science.

Man and his Dwelling Place

History remembers James Hinton as a successful surgeon and author of books and articles on physiology and ethics. A gifted thinker and communicator, Hinton was well placed to address the relationship between science and religion in an age when the two were pitted against each other. First published in 1859, the same year as the Origin of Species, Man and His Dwelling Place takes an ambitiously broad view of the human condition, addressing difficult topics from science, religion, philosophy and ethics. Hinton's arguments against outdated ways of thinking and his approach to human nature were revolutionary, and he took pains to address readers' doubts in a series of question-and-answer dialogues at the end of the book. Hinton's impassioned plea for a bolder spirit of enquiry to better interpret human existence assures this book an important place in the history of science and the understanding of Darwin's intellectual context.

Man and his Dwelling Place

An Essay towards the Interpretation of Nature

James Hinton

Cambridge
University Press

CAMBRIDGE UNIVERSITY PRESS

Cambridge New York Melbourne Madrid Cape Town Singapore São Paolo Delhi

Published in the United States of America by Cambridge University Press, New York

www.cambridge.org
Information on this title: www.cambridge.org/9781108001236

This edition first published 1859
This digitally printed version 2009

ISBN 978-1-108-00123-6

MAN

AND HIS DWELLING PLACE.

MAN

AND HIS DWELLING PLACE.

AN ESSAY

TOWARDS THE INTERPRETATION OF NATURE.

"As for the possibility, they are ill discoverers who think there is no land when they can see nothing but sea."—LORD BACON. *On the Advancement of Learning.*

LONDON:
JOHN W. PARKER AND SON, WEST STRAND.
1859.

There was an old man who had abundance of gold. And the sound of the gold was pleasant to his ears, and his eye delighted in its brightness. By day he thought of gold, and his dreams were of gold by night. His hands were full of gold, and he rejoiced in the multitude of his chests. But he was faint with hunger, and his trembling limbs shivered beneath his rags. No kind hand ministered to him, nor cheerful voices made music in his home.

And there came a child to the old man, and said : Father, I have found a secret. We are rich. You shall not be hungry and miserable any more. Gold will buy all things. Then the old man was wroth and said : Would you take from me my gold ?

Eastern Parables.

CONTENTS.

BOOK III.

OF RELIGION.

BOOK IV.

OF ETHICS.

BOOK V.

DIALOGUES.

MAN AND HIS DWELLING PLACE.

INTRODUCTION.

He who has seen obscurities which appeared impenetrable in physical and mathematical science suddenly dispelled, and the most barren and unpromising fields of inquiry converted, as if by inspiration, into rich and inexhaustible springs of knowledge and power on a simple change of our point of view, or by merely bringing to bear upon them some principle which it never occurred before to try, will surely be the very last to acquiesce in any dispiriting prospects of either the present or future destinies of mankind.

Sir J. Herschel: *Disc. on Nat. Philosophy.*

IT has been well observed that the child and the savage invent an explanation of every thing they do not understand, whilst the man whose powers are matured and disciplined investigates. He has learnt to be patient, and to wait for grounds of knowledge before he supposes himself to know. Thus progress is made. From investigation comes discovery. Our partial and incompetent reason, brought into contact with the great facts of nature, becomes itself enlarged. For the natural suppositions by which man explains the unknown are not equal to the scope of things. They express himself, his ignorance, his limited relations.

All advance in knowledge is a deliverance of man from himself. Slowly and painfully he learns that he is not the measure of truth, that the fact may be very different from the appearance to him. The lesson is hard, but the reward is great. So he escapes from illusion and error, from ignorance and failure. Directing his thoughts and energies no more according to his own impressions, but according to the truth of things, he finds himself in possession of an unimaginable power alike of understanding and of acting. To a truly marvellous extent he is the lord of nature.

But the conditions of this lordship are inexorable. They are the surrender of prepossessions, the abandonment of assumptions, the confession of ignorance: the open eye and humble heart. Hence in all passing from error to truth we learn something respecting ourselves, as well as respecting the object of our study. Simultaneously with our better knowledge we recognise the reason of our ignorance, and perceive what defect on our part has caused us to think wrongly.

Either the world is such as it appears to us, or it is not. If it be not, there must be some condition affecting ourselves which modifies the impression that we receive from it. And this condition must be operative on all mankind: it must relate to man as a whole rather than to individual men.

So far as we could judge without reference to experience, either of these cases might be supposed.

There is perhaps no sufficient reason, *a priori*, why we should not imagine that the appearance might correspond with the fact of things; and, on the other hand, we know that circumstances which affect ourselves do continually modify our perception of objects, so that their appearance differs more or less considerably from that which they truly are. And in some cases this difference of the appearance from the fact is very great. Perhaps nothing can be more unlike the planets than the appearance they present to us; or, to make the case more striking, let us imagine our own earth viewed from one of the other planets. Can any thing be more different from this dark solid varied mass than the bright spot it would appear? The dissimilarity is extreme.

Therefore, when we approach inquiries relating to nature, and the true relations which we bear to the universe, we must be treading on unsafe ground if we assume, without investigation, one of these possible events to be true to the exclusion of the other. We cannot be sure that the world does not differ in extreme degree from its appearance. All experience combines to teach us caution. The history of human error is a history of the taking it for granted that things are as they appear.

Speaking generally, we may say that all speculation hitherto has been based upon the supposition that the appearance of the world does correspond with the fact. All systems are attempts to represent the order of things on that natural supposition. And not only is this the case with philosophical systems, it is equally true of the ordinary and unregulated ideas which lie in every man's mind. All our conceptions are based on the implied postulate that the world is as it appears.

How far the result is satisfactory each man must judge for himself. But it should not be forgotten that another course is open. If we could recognise any element in our condition that should have the effect of causing the appearance of the world in which we are to differ from the fact, the issue of our speculative labours might at least be different from that which it is at present.

That appearances should be deceptive has an evident necessity in the nature of things. For the appearance of every object, or the way in which it primarily impresses us, depends upon our relations in respect to it.* But these relations, infinitely varied as they are, must be ascertained by the study of those objects themselves. We have not any natural or intuitive knowledge of them. Therefore as our relations to the world become more widely known to us we are constantly learning to recognise as the cause of our perceptions facts which are widely different from that which those perceptions at first suggest. Nor do we feel in doing this any embarrassment or difficulty: it is the very thing

* Apparent size, for example, depends upon our distance.

which gives to our conceptions clearness and simplicity.

For right knowledge, it is necessary that the relations between ourselves and the objects that affect us should be clearly understood: that we should know why, the fact being as it is, the appearance must be such as it is to us. The planets appear so small because of our distance, so bright because of the laws of reflection of light: they appear to be revolving around the earth because we are being moved. Knowing these things, it is no longer strange to us to think of those specks of light as orbs kindred to our own: or of the stars so like them in respect to sense as yet vaster worlds glowing with a radiance of their own. We entirely mistake if we imagine that there is any difficulty to the human mind in recognising under any sensuous appearance a fact how unlike soever to that appearance. Nothing is more natural: to nothing is our native tendency more strong. The discovery of facts beneath appearances is the very work of the intellect, and is indeed but the recognition of our own relations to the universe. But there is always a difficulty in first taking this step: that which, when it is familiar, it seems impossible to doubt, when it was new seemed not less impossible to believe. The source of this difficulty lies in our very constitution. For we necessarily think that an appearance corresponds to the fact until by increasing knowledge we have learnt otherwise.

The intellect demands that every appearance should be accounted for. Every impression on us must have some cause; and we necessarily suppose a cause correspondent to every such impression until some other fact be shown to which it may be more reasonably referred. This constitutes the formation of hypotheses; which are accordingly necessities of our mental being. For example, before astronomy was understood, men necessarily supposed that there existed in the heavens a small bright disc such as the moon appears. This was a *hypothesis,* which the recognition of the true moon sets aside.

Hence arises one chief difficulty in the advance of knowledge. For it is the proper work of the intellect in removing ignorance to connect our impressions or sensations with facts different from those which are most naturally suggested to us. The advance of knowledge consists in the substitution of accurate conceptions for natural ones. New truths, therefore, always come, not only with an aspect of strangeness, but in apparent opposition to received and established beliefs; sometimes in opposition to views held sacred, or fundamental to all knowledge. The hypothesis, or cause that had been supposed in ignorance in order to account for the appearance, has a hold upon the mind as if it were a fact certainly known. It is the hardest thing possible for men to remember that such hypothesis has no foundation except their own ignorance. The fact that they have been obliged to suppose it, and that to have denied it without showing how the impressions of which they are conscious could be otherwise produced, would have been to leave a ridiculous vacancy, and to run in the face of common sense, often overpowers all other considerations. The demand upon them to give up that which they have considered as of all things the most certain, is too much. Evidence is of little avail against that feeling. The utmost simplicity, beauty, and necessity in the new opinion often go for nothing in comparison with it.

And there is, besides, always this argument in favour of a hypothesis that has by long use become established as a truth: it is so natural; it answers so exactly to the impression or appearance which it is used to account for. This must be the case; being invented for the very purpose of accounting for our impressions, a hypothesis cannot be wanting in exact correspondence with them. In this respect it must have an advantage, and a very powerful one in relation to some of our strongest feelings, over the truth which seeks to supplant it. For that truth demands reflection and thought; it is in a certain sense opposed to our first natural conceptions, and involves an exercise of reason and a regard to the mutual bearing of various facts. Hence the struggle for the life of a hypothesis is the more prolonged. If the hypothesis be assumed, everything is simple, our impressions need no correcting, and the case is just as it seems. To all this there is nothing to be opposed but the argument that, plausible as that belief may be, investigation and a just use of our powers forbid us to rest in it. The weak part of a hypothesis is not that it does not perfectly account for our impressions, this it can hardly fail to do, but that it will not bear investigation. The existence of that which is seen in spectral illusions or in dreams would account perfectly for their occurrence, and we do indeed at first always account for them so. That is the natural hypothesis; but examination proves it impossible, and we have learnt accordingly to assign them to other causes. Which causes, it may be observed, are very far from being such as we should have thought likely.

These are in part the reasons which render the establishment of a new truth so difficult. Every such truth has to encounter a hypothesis which perfectly accounts for the appearances, makes little demand on the thoughtfulness and reason of men, and, above all, is established as a certain and unassailable truth, based on an experience which cannot deceive. It is no wonder that under these circumstances false views of nature should have struggled long with advancing knowledge. We should not complain that it has been so: that were to find fault with the very faculties and mental tendencies through which alone we have been made capable of learning.

Especially we should avoid the injustice with

which it is too customary to treat the past. We are apt to think that the men who strove so long against opinions which are to us almost self-evident must have been less open to conviction and less willing to abide by the results of investigation than ourselves. But herein we do a twofold wrong: we cast undeserved reproach upon the dead, and inflict a deeper injury upon ourselves. Reading history so, making it feed our own self-confidence and pride, is sadly to abuse its lesson. Men do not alter: in these days they are no more willing to give up what they consider settled facts and principles than they were of old. In all ages men have been willing to apply principles that have been proved true, to do again in other forms that which has been done before; in no age willing, or likely to be willing, to do more. In the past we may read the present: we forget what those men whose errors we pity were called upon to do; we forget how much we owe them for what they did. They were called upon to set aside the very principles on which their mental life was moulded, to abandon as false convictions which seemed to carry away with them the entire basis on which a sound judgment or a steadfast faith could be sustained. And they did it. Trusting in God, the world has given up over and over again well nigh all its most assured convictions; trusting in God that the fact must be better than their thought. Is it for us to boast ourselves; are we willing to do as much again?

The truth is that every generation of men thinks that it has at last arrived at the ultimate principles of knowledge, and that whatever mental revolutions may have been necessary before, no more will be needed thereafter. It must be so. The very fact of men honestly striving to do their best involves it. Man cannot foresee the future; his little horizon must seem to include the scope of heaven and earth. Ever, therefore, he is anxious to know more in accordance with his own ideas, but he cannot anticipate conceiving differently. Yet it might not be impossible to draw from history a lesson that should make us truly wiser, if we would remember that the thing which has been is the criterion of that which is likely to be; and that, as other ages, so we also might be called upon to admit ourselves in error in some of those opinions in respect to which we have been most sure that we are right.

The idea which is commonly entertained of nature is the best conception that men have been able to form respecting it, in the absence of definite, or at least of complete knowledge. Accordingly it corresponds precisely to their first natural impressions, which indeed it is constructed to represent as closely as possible. It is therefore conformable to all experience that the advance of knowledge should bring men into collision with this conception, and that it should exist as an obstacle to a truer interpretation of the facts. If it be the case that our natural impressions fall short of the truth, then, of necessity the ideas to which we have had recourse to account for those impressions must be inadequate. They must embody our ignorance, and differ essentially from those which we should form if the true relations which exist between ourselves and the world were known to us. In a word our conception of nature is a hypothesis.

Like other hypotheses, however, it has had its necessity and its use, nor can it be set aside until the truth be known—the fact itself, and the reason that we are affected by it as we are. The question which demands an answer in respect to the world is at least susceptible of a distinct and explicit statement. We require such a knowledge of our own relation to the fact that truly exists as shall enable us to understand how that fact, being such as it is, should affect us as it does.

Many questions of an abstract nature suggest themselves here. Volumes have been occupied in discussing whether such knowledge be possible; the nature of perception and of consciousness. But the sole answer that will be attempted now is a practical one: for the question is one that must be solved by experience and not by anticipation. It is submitted that man's relation to the fact of the universe may be ascertained by investigation, and that when that relation is understood it may be known also what that fact must be, and why it affects man as it does: and that this knowledge is obtained through thinking more humbly of ourselves; through giving up our natural self-assertion, and being willing to admit that man may be wanting in that which he most confidently assumes that he possesses.

A brief outline of the view that will be advocated is here subjoined. It is thrown into the form of propositions or theses, as a statement of that which is afterwards to be discussed. This plan has been adopted in order that the conception may be presented in its connexion as a whole before any part is treated in detail.

Briefly, the position maintained is this: That the study of nature leads to the conclusion that there is a DEFECTIVENESS in man which modifies his perception; that the universe is not truly correspondent to his impressions, but is of a more perfect and higher kind.

To judge rightly of nature, therefore, we must not be guided by our own impressions merely, but must remember man's defectiveness. For if man be defective, his apprehension and feeling of nature will be inadequate, and that which he feels to exist will differ from the true reality by defect.

Whether this simple change in our point of view, the application of the principle of considering the defectiveness of man in our judgment of nature, have the power of 'dispelling obscurities which have appeared impenetrable, and converting an unpromising field of inquiry into a rich spring of knowledge and power,' may appear hereafter. It has an immediate bearing, thus :

1. Nature (or the universe or the world), is not truly and in itself such as it is to man's feeling. That which man feels to be differs from that which is, apart from him, *by defect.*

We perceive the world as possessing certain qualities, or as existing in a certain way which we call physical. We term it the *physical* world.

This mode of existence involves inertness. That which is physical does not act, except passively, as it is acted upon. Inertness is inaction.*

That which is inert, therefore, differs from that which is not inert by *defect,* (by absence of action or of active power.)

2. We cannot avoid conceiving another mode of existence besides that which is inert. We conceive of Being which possesses a true spontaneous and primary activity. This is necessary, since there must be such a true activity, or there could not be any action at all.

To this truly active mode of Being the word *spiritual* has been applied; and in this sense that word will here be used. That to which inertness does not belong, but which truly acts in a way in which physical things do not act, is meant by the term spiritual.

* From the Greek word ερδω, *to act.*

The physical therefore, differs from the spiritual (in this particular of its inertness) *by defect.*

3. It is submitted that it is man's defectiveness which makes him feel the world as thus defective: that nature is not truly inert, but is so to man's feeling by defect in him.

We have conceived nature to be inert, or physical; man to be not inert, or spiritual.

It is submitted that investigation demands that we should correct this natural supposition: That the perceived inertness or defect in nature is due to man's defectiveness.

4. Either the universe is defective as being without action (inert), or man is defective. There is *to us* an inertness, it determines our whole state. We have to learn whether it be man's or nature's.

SCIENCE gives answer to this question. By it proof is given that the perceived defect must be ascribed to man's condition, and that nature is not truly inert as it is felt to be. His own condition having imposed on man a false opinion respecting the universe, science emancipates him therefrom; it brings man face to face with nature, and makes him know *himself.*

5. The history of science is the attempt of man to understand the universe on the supposition that the inertness (or defect) exists in nature as it appears to him to exist. But this attempt leads to the result, entirely unforeseen, of transferring the

defect to himself; and proving that both the fact of nature and his own state of being are different from that which he supposed.

This result science accomplishes—

1st. By demonstrating an absolute inertness in that which appears, bringing all *phenomena* under the law of passive or physical causation.

2nd. By giving evidence of a fact different from that which appears to us, showing that it deals only with phenomena, and not with the very essence of nature.

It is affirmed, therefore, that inertness does not belong to the essence and true being of nature,* but only to the phenomenon.

It is introduced by man. He perceives defect without him only because there is defect within him.

6. To be inert has the same meaning as to be dead. So we speak of nature, thinking it to be inert, as 'dead matter.' To say that man introduces inertness into nature, implies a deadness in him: it is to say that he wants life. This is the proposition which is affirmed. This condition which we call our life is not the true life of man.

7. The book that has had greater influence upon the world than all others differs from all others in affirming that man wants life, and in making that statement the basis of all that it

* The proof is deferred, not belonging to this place. See Book I. Chap. i.

contains respecting the past and present and future of mankind.

Science thus pays homage to the Bible. What that book has declared as if with authority so long ago, she has at last deciphered on the page of nature. This is not man's true life.

It is a willing homage. For all men love the Bible: some of those not least who have most felt themselves compelled to oppose it. In every heart the love is deeper than the hatred. For what book has sounded so the depths of experience, or scaled like it the highest pinnacles of thought? What man has not learnt through it better to know himself?

Therefore if the thought that man wants life seem at first strange to the intellectual apprehension, the conscience and the heart respond. This is not our true life. Illusion, and disappointment, and wrong are in it. We ought to be other than we are.

8. The statements of the New Testament respecting the course and history of the world, starting with a deadness in men, end in their being made alive.

We naturally conceive the world to be the scene of man's probation. The Bible represents it as the scene of his redemption. Man is being made alive: rewards and punishment, threatenings and retribution, take their place within and in subordination to this end.

9. That man wants life, means that the true life

of man is of another kind from this. It corresponds to that true, absolute Being which he, as he now is, cannot know.

He cannot know it because he is out of relation with it. This is his deadness. To know it is to have life.

10. To that absolute fact of Being the Bible applies the words spiritual and eternal. These are the right words. To be spiritual is to be not inert. To be eternal is to BE.

The unknown fact of nature is the spiritual and eternal world; 'the things that are not seen.' But man wants that true life which would place him in union with it. Therefore to him the world is temporal and physical. He does not know the fact. Therefore he feels that to be which is not.

In other words: there is not a physical world, and a spiritual world besides, but the spiritual world which alone IS, is physical to man: the physical being the mode in which man, by his defectiveness, perceives the spiritual. We feel a physical world to be; that which IS, is the spiritual world.

The necessary bearings of the conception that has been thus proposed may relieve from the charge of presumption the attempt to comprehend in one view so many things as are included in this volume. That is a task imposed by the nature of the case.

The same remark will apply to the objection that

will be felt to the mingling of science and religion. The justification of this proceeding is simply that it is believed to be right.

It may be that the separation of our thoughts concerning things physical and things divine is a disunion of our being, a partition into two imperfect halves of that which rightly constitutes a harmonious whole. The separation has indeed been needful and eminently useful, but only as a temporary expedient; it cannot be a permanent relation. Religion will not unite with a science based on the supposition that man is living and the universe dead, but a science that recognises deadness in man in that very act becomes religious. Science is of necessity divorced from religion while it rests in phenomena, but when it takes cognizance of man's relation to that which is not phenomenal, it is reunited to its source.

The union of science and religion is not optional; a thing which may be attempted or avoided. That union is a fact, to which we must conform ourselves. Science is religious. All things are so. There is no object of human activity or interest of which the same thing may not be said. Nothing is unreligious but by error and ignorance: only so long as we do not see what it is, and for what purpose it exists, can any form of activity or of thought be kept apart from our religious life.

For religion is simply that which concerns the very fact and reality of our being. That which constitutes anything religious is its being brought into relation with that fact, and placed in its true bearings. That is religious which is felt and known aright, in its own true nature, and not according to the mere appearance to ourselves.

Religion is the one thing in which all men are interested; the one absorbing inquiry to which no man is indifferent. What am I, what is the world? Why am I here, and what will be the result? What justice, what love, what rightness, what hope, what end? These are questions which no man ceases to ask, or will cease. To these questions if any man give answer, the world listens with credulous and eager ear.

But other interests are partial, and limited, and transient. They ruffle the surface of our life, but do not stir its depths. Men make them the objects of their devotion, and try to be content, and fill out their emptiness with pomp of words and specious self-congratulation, because they fail in their attempts to deal with those deeper and dearer questions which rack their souls in secret.

BOOK I.

OF SCIENCE.

Nature is the domain of liberty.—*Cosmos.*

CHAPTER I.

OF THE WORK OF SCIENCE.

'Tis life of which our veins are scant.—*The Two Voices.*

PROOF is of three kinds: First, the Logical, which rests on premises, and demonstrates that according to the laws of the human mind a certain conclusion follows.

2nd. The Historical, which shows that if the case be as affirmed, the course of human thought in relation to it must have been such as it has been. It accounts for the rise and progress of opinion.

3rd. That which might be called the Expository, which, taking the phenomena as they appear, gives a simple statement of the fact which carries its own conviction. Such is the evidence on which the Copernican astronomy is received by the mass of educated men.

Each of these modes of proof is indispensable; but they are by no means of equal authority. The logical is principally useful as a means for advancing knowledge. Its conclusions can never have more certainty than the premises, and its end is chiefly to free us from false ideas by leading us to false results when we reason from them. It makes the latent error manifest. Logic has less to do with that which is true than with that which it is reasonable for us to think with our particular amount of knowledge. The historical and expository proof have more positive value. The light which they throw upon that which has been, and which is, gives them an authority to a certain degree independent of ourselves.

The argument from premises to conclusions will be the least employed here, not because it is inapplicable, but because it is the least appropriate. It neither can nor should produce conviction. If an improbable conclusion be enforced by such reasoning, the premises are immediately suspected, and rightly so. It will be sought, rather, to unfold the conception that man is such as he is by a want of his true and perfect being, and that he is being raised from this state by having the true life imparted to him; and so to exhibit this conception in its relation to the facts of human life that it shall be felt to be the solution of the problem of humanity, the true interpretation of history, the key both to what men have thought and what they are.

If it can be made manifest that the deadness and redemption of man is the reconciliation of all enmities, the oneness of all opposites; that it demands of no man that he should abandon that which he has revered as sacred or valued as true, but is rather the perfecting of all these things; that it demands a willingness not to give up, but only to receive, putting new meaning into our habitual words, new life into our daily work, and making light to be where darkness has been: this is the evidence on which reliance will be placed.

Two or three observations will serve to guard against some possible sources of misapprehension.

1. The first of these relates to the nature of Language. Words necessarily express to all persons their own conceptions. Hence the difficulty of conveying by them ideas that are new, even in any branch of ordinary knowledge. Much greater is

this difficulty when the question relates, as now, to the entire conception of existence. No word can be used that is not already fixed as it were to a different class of ideas, so that in its new use it may either fail to convey the meaning or seem to be misapplied. This difficulty is inherent in the subject, and is certainly much increased by want of skill on the writer's part. Perhaps, however, it will not be found greater than any one who will seek for the meaning, and make allowance for deficiencies in respect to words, whether unavoidable or inadvertent, will easily surmount.

In no respect does greater embarrassment arise from words than from the various use of the word *To be:* employed as it is to express either true existence or mere appearance: *absolute*, as it is termed, and *relative*. We say of God, He IS; but we use the same word of a shadow, of which the essence is that there *is not* light. The being of a shadow is only an absence, yet we cannot mark this by the words which express existence. We cannot deny that a shadow 'exists.' It exists as a shadow, or has such existence as a shadow has. We say there *is* darkness, so expressing negation or denial. This source of error must be remembered and watched against; it cannot, in the present state of language, be avoided.

'It is a rule,' says that great master of discourse, Lord Bacon, 'that whatsoever science is not consonant to presuppositions must pray in aid of similitudes. For those whose conceptions are different from popular opinions have a double labour, the one to make themselves conceived, and the other to prove and demonstrate; so that it is of necessity with them to have recourse to similitudes and translations to express themselves.' The use, therefore, of illustrations and comparisons drawn from sensuous things, in the following pages, will not be thought inappropriate, or designed to snatch an assent from the fancy which the calmer judgment should withhold. Nor will the points of difference which must exist in all similitudes from that which they are used to illustrate be held to imply an attempt to argue from one thing to another, disregarding the diversity of the cases. The similitudes are used to aid the conception of the thought.

2. That man wants life may seem to exclude individual responsibility. Certainly no opinion can be true that sets aside the moral instincts and does violence to the conscience. It may suffice here to state that our actions, in so far as they are our own, are held to be not necessary, and that we are, therefore, responsible for them. The doctrine of man's deadness, so far from diminishing, strengthens and renders more profound the sense of sin.

3. It may seem unnatural to speak of a conscious existence as a state of death. But what is affirmed is, that a sensational existence, such as ours, is not the Life of MAN; that a consciousness of physical life does itself imply a deadness. The affirmations—that we are living men, and that man has not true and absolute Life—are not opposed. Life is a relative term. Our possession of a conscious life in relation to the things that we feel around us is itself the evidence of Man's defect of Life in a higher and truer sense.

Let a similitude make the thought more clear. Are not we, as individuals, at rest, steadfast in space; evidently so to our own consciousness, demonstrably so in relation to the objects around us? But is man at rest in space? By no means. We are all partakers of a motion. Nay, if we were truly at rest we could not have this relative steadfastness, we should not be at rest to the things around us, they would fleet and slip away. Our relative rest, and consciousness of steadfastness, depend upon our being not at rest. These are moving things, to which he only can be steadfast who is moving too. Even such is the life of which we have consciousness. We have a life in relation to these physical things, because man wants life. True life in man would alter his relation to them. They could not be the realities any more, he could not have a life in them. As rest to moving things is not truly rest, but motion; so life to inert things is not truly life, but deadness.

How should this be otherwise than a familiar thought to those who have taught us that man, as he now is, cannot know the absolute, that he deals only with *phenomena*, and not with the very fact of being?

The word phenomenon has been introduced into science to denote that the true essence of nature is different from that which we can know by sense or conceive by intellect. The things which we perceive or think, do not correspond to the very fact of being; that is unknown. Phenomena are appearances.*

But here a difficulty arises. For if these things which we know be but phenomena, then do we feel them wrongly. To us they are the realities, they determine our whole life and condition. Our life is a life in that which is but phenomenal. Thus science demonstrates man's defect. Our perception and feeling are not true. We lie under illusions which have relation not to our intellect alone, but

* From the Greek, φαινομαι, *to appear*.

to our very being. We cannot separate them from ourselves. While man is such as he is, that which can only be appearance must be reality to him. He feels himself in conscious relation only with that which is not the very essence, the truth of things. Stripped of technicality, this is the plain teaching of our science. It may interpret to us our secret discontent, and explain the wearisome failure of our lives.

It cannot be otherwise than that, if the true essence of nature be unknown, we should have been compelled to think it to be such as it is not. For we cannot think there is no fact in nature. The reality impresses itself upon us too strongly. Nature IS. It is no illusion that we are in a real, actual world. So that there has been no escape. Compelled on the one hand to be sure that the universe has a true existence, and on the other unable to know that to which this true existence belongs, the issue has been unavoidable; we must have been under illusion, and have believed that to be which is not. This last fruit of investigation, the discovery that man cannot know by sense or intellect the true 'being' of nature, brings into harmony all the various thoughts of men, and shows to what end they have been working.

For men could not understand that this was their true relation to nature, until through long and varied experience the conviction had been forced upon them. It is natural to men to suppose, and indeed to feel absolutely sure, that they do know the true fact of nature, and that it is such as it appears. That it should be so is indeed implied in the meaning of the word 'appear.' That alone is an appearance which men naturally suppose to be the fact until they have learnt otherwise.

And the more must it have been difficult for men to recognise this truth, because of the bearing it has upon themselves; for the phenomena that sense perceives, and that science investigates, are the realities of their life. If they are not truly the reality, then must their own being be defective in a sense they are not prepared to admit. They are under illusion in such a way as must entirely alter their own conception of themselves. The true being of man cannot be in them. The true being of nature is hidden from our eyes because there is not that within which answers to it.

It is a remarkable thing that men should have rested in the assertion that we cannot know the essential being of nature, without recognising that this fact necessarily places us under illusion, and causes us to attribute being to that which does not possess it. We understand, however, why they should have failed to perceive this evident consequence of their position when we see what it involves. That man has not his true life, must have taken him long to learn. All our prepossessions, all our natural convictions, are opposed to that belief. If these activities, these powers, these capacities of enjoyment and of suffering, this consciousness of free will, this command of the material world, be not life, what is life? What more do we want to make us truly man? This is the feeling that has held men captive, and biassed all their thoughts so that they could not perceive what they themselves were saying.

Yet the sad undercurrent has belied the boast. From all ages and all lands the cry of anguish, the prayer for life unconscious of itself, has gone up to heaven. In groans and curses, in despair and cruel rage, man pours out his secret to the universe:

writing it in blood, and lust, and savage wrong, on the fair bosom of the earth; he alone not knowing what he does. If this be the life of man, what is his death?

That the true 'being' of nature is not inert rests on a threefold argument. We feel that the phenomenon is inert, and controlled by passive necessity. The question is, therefore, whether this feeling on our part corresponds to the truth of nature as it IS: whether that which exists, apart from man, be thus inert, or our feeling be due to man's own state of being. In other words, whether inertness be not one of the respects in which the phenomenon differs from the true being of the universe.

1. Inertness necessarily belongs to all phenomena. That which is only *felt* to be, and does not truly or absolutely exist, must have the character of inaction. It must be felt as passive. A phenomenon must be inert *because it is a phenomenon*. We cannot argue from inertness in that which appears to inertness in that which IS. Of whatsoever kind the very essence of nature may be, if it be unknown, the phenomenon must be equally inert. We have no ground, therefore, in the inertness we feel, for affirming of nature that it *is* inert. We must feel it so, by virtue of our known relation to it, as not perceiving its essence.

2. The question therefore rests entirely upon its own evidence. Since we have no reason from the inertness of the 'phenomenal' for inferring the inertness of the 'essential,' can we know whether that essential be inert or not? We can know. Inertness, as being absolute inaction, cannot belong to that which truly is. Being and absolute inaction

are contraries. Inertness, therefore, must be a property by which the phenomenal differs from the essential or absolute.

3. Again, Nature does act: it acts upon us, or we could not perceive at all. The true being of nature is active therefore. That we feel it otherwise shows that we do not feel it as it is. We must look for the source of nature's apparent or felt inertness in man's condition. Never should man have thought to judge of nature without remembering his own defectiveness.*

What can be more simple than that our own state should affect our feeling, and have necessitated our thinking of the world as it is not? Universally the principle is recognised in respect to individual things, that our own condition affects our feeling, and that we must have regard to that condition as an element in judging. The application of this principle to the investigation of the world as a whole, and to conditions affecting all mankind, is all that is contended for. Before we can know by our feeling of the world what it truly is, we must understand man's condition in relation to it.†

Here is the especial work of Science. By inves-

---

* The perplexity that is always felt respecting perception, and man's conscious relation to the physical world, whenever the question is agitated, arises in great part from the incongruous supposition that *inert* things *act* upon us. It is evident that in whatever world man might be, *if he were conscious only of phenomena*, that is, if he did not perceive the essential being of it, he must be conscious of being in an inert world. That condition carries with it the conscious perception of inertness unavoidably. We may not, therefore, assume any other cause for our perception of nature as inert.

† Guarding against what Bacon calls, 'the Idols of the tribe;' or those errors respecting nature to which we are prone through circumstances which affect all men.

tigation of that which he feels to be, man learns his own condition, and becomes able to interpret the appearance of the world. This is its end and use, the part which it plays in the great work of human life. For men, pursuing their own ends, fulfil God's. All human activity bears witness to a larger purpose in it than any that is consciously present to the worker; often the object sought being of little value compared with the result that is achieved.

For many generations, now, the chief energies of thinking men have been devoted to physical research. Unwearied has been the diligence, patient and self-sacrificing the toil, that have been brought to the task; glorious the offerings of self-denial, enthusiasm, life, that have been laid upon that altar. The results may look cold, barely set forth by weight and measure, or clothed in uncouth formulas, but a warm life glows beneath. The dark crater is instinct with fire. For those results the largest hearts of human mould have poured themselves in passionate fervour upon nature, and ecstasies of joy and hope have thrilled to weakness frames which no labours could exhaust; for God had moved them. The wonder of His works was as a spell upon them; the mystery and beauty of the universe wrought like a command within. They stretched forth their hands unto the Infinite.

And what have they grasped? Some mathematical relations, some undefined ideas about forces —a perception merely of undeviating law? Have they but inaugurated a ceaseless strife between the emotions and the intellect—an everlasting protest of piety against conviction? Must they content themselves with physical advance, and take refuge from perplexity of heart in bridging oceans and annihilating space, the bright visions which lured them on fading like the enchantments of a dream? Is this the end?

Not so. In creating science men have done more than they knew. They have prepared the way for the removal of an illusion. Hence the discord. For the truths of science will not blend with the conceptions we have formed of nature without a shock to ineradicable feelings. The great thought of science is necessity; the human soul demands above all things freedom, not only for itself, but even more for the Power by which the world is governed. Therefore it is that science and religion have been at strife. Our conception of nature as inert would not permit it to be otherwise. The establishment of a necessary connection between natural phenomena has seemed to put a chain upon the hand of God, and substitute a dead mechanism for the living sympathy that men had found in wave and mountain, in storm and sunshine, in the beauty of the earth and glory of the sky. The present state of science in the minds of religious men is, for the most part, a result of opposing forces, a compromise between the ideas of physical causation and of the direct action of the Creator. But it need not continue so. This is an embarrassment which arises in the course of advancing knowledge, but which ceases with the misapprehension from which it springs. The work of science is to rectify our thought of nature; to show us that the deadness we perceive is but our own. Here is the reward of all that patient toil; the rightful fruit of that prolonged investigation. The explanation of the universe which ignorance has supposed yields to a juster knowledge: there is not a defect in nature, but a want of life in man.

For men have naturally believed that the sole result of Science would be to enlarge our knowledge of that which appears; to discover the relations of phenomena, and give us control over physical things. But Science has an evident adaptation to do more than this; to make us know ourselves more truly, to reveal to us not only that which is without, but that which is within. Thus it places us in an altogether different attitude in respect to our knowledge of nature, enabling us to attribute to its true source the defect we feel.

There are three words in established use: *appearance*, *phenomenon*, and *fact*. Between appearances and facts there is the widest distinction, they are even opposites. Yet the word phenomenon is used sometimes for one, sometimes for the other. Confusion of thought must result from such a use of words; for things that are equal to the same ought to be equal to one another.* But the reason of this vacillating language is, that while men are compelled to say that phenomena are but appearances, they do truly think of them as facts, or realities; for they are felt by us as real. We speak of things in one way and think of them in another, for we can only truly think of phenomena as but phenomena, by constantly remembering man's want of life.

But the right use of these words is distinct and simple. Appearance is that which is to our sense, but is not true to our thought; *e.g.* the appearance of the moon is a bright disc. Phenomenon is that

* Coleridge notices 'the unconscious irony with which the same things are termed indifferently facts and phenomena.'

which is to our thought corrected and checked by our senses; the moon itself therefore is a phenomenon: it is that which we can think. Fact is that which truly and absolutely exists: the essential BEING of nature which we cannot think. Fact is that which IS: Phenomenon is that which is to our conception: Appearance is that which is to our sense. Defect of knowledge makes appearances facts to us; defect of being makes phenomena facts to us. In the true life of man, the fact alone should be the fact to him, and phenomena should be but phenomena instead of being, as now, the realities of his existence: even as true knowledge is to know and feel appearances to be but appearances, instead of their being, as they are to ignorance, realities. Now, the phenomenon is real to us, moulds and determines all our experience. We express this fact by saying we are the slaves of matter. The discordance of our state with the aspirations and unquenchable assertions of our soul is felt, but not understood. It is want of life in man under which we labour, that makes the universe physical to us, and subjects us to the tyranny of inert necessities. For nature IS not as we feel it. Thus do we perceive and feel another different fact, thus to feel it not for ever. Life is to be given to man, a life whereby, being more, he shall feel more truly. The instincts which assert for man a truer, worthier being may assume a loftier tone. Science is their friend and servant, not their enemy; revealing deadness in respect to man, it explains the mystery of his present state, adds emphasis to the prophecy of a different future. Man shall be made alive; altered not in his circumstances, but in himself. The physical testifies of the spiritual; the dead defective world, of which we are conscious, tells us of man, of his deadness, of his need to be made more. 'Signal the happiness of humanity to be environed with an imagery so resplendent; but happier still, that amid the weaknesses, the very ashes of our being, the power remains to apprehend that imagery as a symbol.'*

---

* NICHOLS' *Architecture of the Heavens.*

CHAPTER II.

OF THE LAWS OF NATURE.

These three be the true stages of knowledge, and they are as the three acclamations, Sancte, Sancte, Sancte! holy in the description or dilatation of his works; holy in the connection or concatenation of them; and holy in the union of them in a perpetual and uniform law.

LORD BACON: *Of the Advancement of Learning.*

WITH true action we necessarily connect moral conceptions. The ideas cannot be dissociated. And that to which moral conceptions apply is by all termed spiritual. For this reason the fact of nature has been affirmed to be the spiritual world. That it is so, follows from the proposition that inertness does not belong to it. The argument would be the same, if there were insuperable difficulties in conceiving how man should be made to perceive the inert phenomenal world by his presence, in a defective state, in the spiritual world. The proof that it must be so would be none the less complete. But this is by no means the case. On the contrary, the light which the facts of our experience receive from the perception that it is truly a spiritual world with which we are in relation, and that it is physical to us only through man's defectiveness—that its being physical to us denotes, and is evidence of, a dead state of man, which else we should not know—is stronger demonstration than any that rests merely on the intellect. For this evidence embraces all our faculties, and appeals to all our being; revealing the source of our inward strife, and taking away perplexities by demonstrating how they must have arisen.

For so we understand why we are in a physical world. Why, although conscious of adaptation (or thinking ourselves so) to a different mode of being, we are so surrounded and hemmed in, subjected to passions and necessities due to physical laws. We are thus because man wants life. A defect having its seat in him is thus felt externally, and seems to constrain him from without.

And thus, too, it is that we have conceived two modes of being: one low, inert, passing; the other higher, free, eternal. This is because man has believed the phenomenal to have true existence, believing his own deadness to be, as he perceives it, in that which is around him. But that our experience is truly due to a spiritual fact, altered to us by man's defect, speaks for itself. We feel that we are in relation with a world that is not inert or dead. Science, in presenting nature as inert, crushes our instincts, baffles our deepest convictions. Man we would willingly grant to be in a dead, lifeless state; but not the universe. It is too full of glory and of beauty—a beauty that is to us sacred, that we cannot but call holy. Nature is bound to us by ties so deep and tender, it is so high above us, stirs us with influences so mysterious, speaks to us in words so moving, sympathizes with us so truly, chides us so gently, so fervently inspires, or sternly warns, holds out to us for ever so bright a pattern, it cannot be the slave of a mere dead necessity. It seems a ruthless hand that tears away so bright a veil, and shows us—not nothing, not a dream, but a dead block, worse than nothing; a cold carved image with unheeding eyes, on which we lavish love in vain; which stamps the life out of our hearts, like idol gods, by blind mechanic impulse, and no more.

Therefore it is that so many men of strong affections and imagination oppose themselves to science. They cannot bear to have all this glory and significance reduced to mere results of physical conditions. All that for which they value nature is destroyed in such explanations. They loathe to think that the tenderness and awe which move them so are but subjective enchantments.

They say, that science does not account for that which they perceive: like the knife of the anatomist, it pursues the life in vain. For to this clear issue the case is brought, man does introduce into nature something from himself; either the inertness, the negative quality, the defect; or the beauty, the meaning, and the glory. Either that whereby the world is noble comes from ourselves, or that whereby it is mean; that which it has, or that which it wants. Can it be doubtful which it is?

The course of nature is constant and unvarying, and this is the ground upon which its inertness is affirmed. For this reason, together with our own consciousness of exertion when we would produce physical changes, we assert the inactivity of nature. If the phenomena were conformed to no laws that we could trace, we should admit nature to be active. But it is evident that invariableness is not proof of inaction. RIGHT ACTION is invariable; RIGHT ACTION is absolutely conformed to law. Why, therefore, should not the secret of nature's invariableness be, not passiveness, but rightness? Rightness of action, being ever one, absolutely unchanging except in form, would appear, if not understood, as an inert necessity. If invariableness implied passiveness, then God himself must be inert, who changes not. Man's deadness and disharmony from nature, which cause him to be variable, self-indulgent, and a transgressor of law, make him believe the fulfilling of law, which he everywhere perceives around him, to be a dead necessity.

Action and necessity in one: this is the fact of nature's undeviating laws—right action, which is necessary where love is.* The fact of love becomes the phenomenon to us of an inert necessity. For love is the true necessity of nature; the

---

* It might be asked, 'Can holiness be predicated of nature as of man? Invariability in the watch I make, is different from invariability in the child I educate. The child has a will and freedom, and his right action I call holiness; but unless the watch also has a will and freedom, I should not call its right action holiness.'

It is not the phenomenon that is spoken of, but that true being, itself unperceived, which causes the phenomenon to be perceived by us. The phenomenon is inert, and therefore not possibly holy; but we speak of the essential fact, which is different therefrom and not inert. Of this, holiness is affirmed, because the phenomenon is uniform. The uniformity in that which we feel to be (by virtue of which alone we can construct a watch), depends on right action, and not on inaction in that which is.

To take an illustration. If when we look through a stereoscope it is said to us that the object is double, we might reply in the same way, 'How can that which I see be said to be double? it is single.' True; it is not that which is consciously present to our perception that is spoken of, but the object which causes us to have such perception, and to think aright of which we must remember the subjective laws of vision. We correct our perception, as it were, by withdrawing our eye from the stereoscope. So we must mentally withdraw our eye to judge of the true reality of nature. Not of that which is consciously present to our perception, but of that which truly is, we want to learn.

necessity by want of which in us comes all our misery: by want of love we know not love in that which is around us. If that moral life, which makes action necessary, had been in man, never would he have inferred inaction from the necessity of nature: never have supposed a deadness from the very proof of life.

What joy it were to know in the unchanging laws, in the ever more widely revealed necessity, of nature, the fact of holiness! And the proof is evident. For the fact of nature cannot be inert. And if there be not inert necessity, why the uniformity? The question admits but one answer. That uniformity is rightness. Rightness must be uniform. Remembering our conscious want, our known defectiveness, all is clear. Deadness can no more be believed in nature. The universe rises to its true level in our regard. A grandeur, awfulness, and joy unspeakable, clothe all our life. It is a sacred thing to live. The secret stirrings and heavings of our hearts, the throbbing passions, the awful questionings, the baffled strife to penetrate the breathing mystery around us: all are seen. The dead humanity embraced in universal life.

That the invariableness of nature bespeaks holiness as its cause doubtless involves an appeal to man's moral sense. If he were merely an intellectual being, there could be no such argument. But the appeal to an inevitable conscious association of right and wrong with true action, the moral distinctions drawn by all, surely has not less weight than an appeal to a perception of intellectual relations. The latter is as much a matter of consciousness as the former. There is the same necessity of belief in the one case as in the other. It is affirmed that the laws of our nature demand, that if we banish inertness from our thought of nature, we should introduce into it the idea of rightness. There must be reason for the invariableness of the phenomena. We are bound by our constitution to attribute rightness to nature if it be not inert.

There is no inert necessity. That only appears, or is felt by us; it is the phenomenon of a different fact. It is easy to understand how man's condition should be such that he should necessarily have conceived unvarying ACTION as inert necessity. It needs only that he should *not* perceive the action.* Where there is true action, not recognised, there passive causes must be supposed. Conceive some being, ignorant of man, ignorant of his spontaneous power of motion: he would necessarily suppose, on observing human movements, that some force operated upon man from without, to produce them. And if they were always the same under the same circumstances, if they were conformed to unvarying principles, to such a being there would be, in human motions, just the uniformity we perceive in nature, just such an appearance of passive law. We conceive of force as existing apart from us only by virtue of our own consciousness. We feel force; it is the phenomenon. The idea of passive forces in nature is an inference solely from our own sensations, and it is a mistaken one. Force implies difficulty, inability; not to say physical sensation like our own. In nature is an absolute, unchanging oneness of fact in ever-changing form. These changes of form impress us with the feeling of force or exertion, when taking place in certain relations

* That is, indeed, only that the essential being of nature should be unknown.

to our consciousness. Force is subjective. As well might we suppose pain in nature as that passive force which we imagine. A similar error we have already escaped from in respect to the sensations of light or sound. Our natural impression is, that the light we perceive exists externally; but we have learnt to recognise a different cause for our perception. We suppose ourselves to be such that an external motion impresses us with a sensation of luminousness. We are such that the ACTION which is in nature impresses us with the feeling of passive force.

Thus the various *forces* which science supposes, as producing the phenomena of nature, are easily understood. They are conceptions necessary for us. They belong to the phenomenon. Science does not affirm them as existing, only as apparent. Taking to herself the position of dealing only with phenomena, she assigns to these forces also but a phenomenal character. The fact of nature is felt by us as a passive existence, subject to these various forces, because we are defective, and do not feel it rightly. The passive forces have been necessarily supposed, because there cannot be true action in that which is but a phenomenon.*

---

* Because that which is a phenomenon cannot EXIST. It has a relative existence only; it 'is to us,' *i. e.*, it is felt by us as existing; that which truly exists being different. Of course, therefore, there cannot be action in a phenomenon, there not being existence. It is of necessity characterized by inertness. It has relative or apparent action, but is in itself and absolutely inactive. To say that the phenomenon is different from that which truly is, and to say that it does not exist, are exactly the same thing. The fact exists, and the phenomenon is felt by us to exist, because of the existence of that fact. Thus it is easy to understand how an *inert existence* has been supposed. The idea is self-contra-

So man's existence in relation with being that is truly active, the fact or essence of it being unknown or unperceived, would necessarily cause him to perceive such inert uniformity as he does. The inertness of nature to our feeling rests only on our not consciously perceiving it as it is. And our feeling nature as inert involves our consciousness of exertion, involves our conscious activity in respect to physical things. Our feelings in respect to the world, as passive, necessarily flow out of its being but phenomenal. In feeling the phenomenal as real, our experience must be such as it is. Consciousness of exertion is inseparable from perception of inertness in nature.

So far we may see clearly. That which we feel to be is inert, because it is not the very essence of that which IS; and the tracing of absolute invariableness and necessity throughout all nature presents to us a wholly different aspect. It no more makes the fact of nature passive, no more sets aside the life. Science has a new attitude to the moral sense. We rejoice to trace necessity in the phenomenon, for that is love and holiness in the fact. To be able to refer all things in nature to invariable laws, to see how everything must necessarily be; this would be

---

dictory, yet it must have arisen. Ignorance necessitates belief in the existence of that which only is to us, *i.e.*, of the phenomenon. But the phenomenon is found to be inert, before it is recognised to be only a phenomenon. It is found to be inert while it is still supposed to BE. Thus the notion of inert existence comes to be entertained, and grows familiar insidiously, so that its impossibility is overlooked; and men who maintain that the things that are to us are but phenomena, and do not truly exist, still regard the fact of nature, which does exist, virtually as inert.

the glory and the joy of man. This would be to demonstrate in nature undeviating holiness, perfect and unfailing love. Take away the inertness, understand that it is due to man, and the reducing all things to an apparently mechanical necessity revolts the soul no more. The necessity is no more mechanical.

The investigation of phenomena does reduce them all within the sphere of merely passive laws. Form only changes, and every change confesses a necessity. Nor does the evidence of this fact demand that all particulars should be discussed. It has demonstration in the nature of the case. Force cannot otherwise exist than under the control of such passive necessity. It is its nature, part of its definition, that it is determined by resistance. Force and resistance are correlative. All natural processes, considered in respect to force, resolve themselves into a passive necessity. Of all things, as the fact is one, one also is the appearance; force obeying necessary laws. Passive change under equally passive resistance or control.*

This is the phenomenon whereof the fact is holiness. Nature is holy, not in figure, not in seeming. In deed and truth the fact of nature's life is holiness; the seeming is necessary passiveness. This is the distinction of the phenomenon from the true and essential fact, that the BEING and therewith the action is wanting. This world is the spiritual world, not known. To be as we are, is to be in the world that truly is, but blind and unper-

* For a further reference to this point, see Book V. Dial. iii.

ceiving, and to have our life, therefore, in a world of mere phenomena, which *is not.* Thus to live is to be under illusion, and spend our days as a dream. This is the unreality, the unsubstantialness of this world, which men's inmost hearts affirm, which has so often found for itself a voice. The world is an illusion, a dream, a mockery. Life deceives us, its promises are lies; it yields no satisfaction, only hope and desire incessantly renewed, a thirst never slaked. That is true. The phenomenon must be unreal, and if we think it to be the true reality, then we are dealing with an unreal world; a world that to be known aright should be to us but a sign of other and higher being, that cannot disappoint.

Because nature is spiritual, science has been compelled to introduce the conception of law. Incongruous as it is with our thought of an inert substance, it has been felt to be not less natural and true to instinct than indispensable for theory. And rightly. In nature law is fulfilled. Perfect obedience is there. For the fulfilling of law is love.

Well may nature bind us with so mysterious a charm, and thrill us with a potency so magical. That Rightness constitutes the deep secret of her being, binds her infinitely close to us, makes her truly ours. No other tie could constitute so true a union. For rightness is the deep secret of our being also. In spite of the evil of our nature, in spite of daily and hourly sin, the strongest passion of humanity is the love of right. Alienated from rectitude, man is at war with his own life. Thence comes our woe, our misery, our sense of loss; being wrong-doers we are alienated from our true and only home. Because she is right, nature is ours: more truly ours than we ourselves. We turn from the inward ruin to the outward glory, and marvel at the contrast. But we need not marvel; it is the difference of life and death; piercing the dimness even of man's darkened sense, jarring upon his fond illusion like waking realities upon a dream. Without is living holiness, within is deathly wrong.

CHAPTER III.

OF THE ILLUSTRATION FROM ASTRONOMY.

CALLICLES.—I know not how it is, Socrates, you appear to me to speak well. Yet that which happens to most, happens to me; I am not quite persuaded by you.—PLATO: *Gorgias.*

THE difficulty which is naturally felt in conceiving that the fact which causes our experience is spiritual may be much diminished by the aid of analogy. Not that analogy furnishes any part of the evidence on which the statement rests. That evidence claims to have a demonstrative basis in science, which demands that we should ascribe the perceived inertness to man, and recognise that the inert phenomenon denotes a fact that cannot be inert. But however sufficient these or any other proofs might be granted to be, there still remains a difficulty in respect to the feeling, a strangeness, and as it were an unnaturalness, that has its seat chiefly in the sense, and which might express itself in such terms as these: 'What does it avail to prove the world not physical? of what use is it to bring arguments that these things which I see and handle, which I use for food or clothing, which are passive before my touch, are spiritual? I know that they are not. This is the world in which I am, and it is unspiritual enough.' All must be subject for a time to this feeling: it is chiefly the result of habit, and soon ceases to cause any embarrassment. It might be sufficient to reply that these things are the *phenomena*, the physicalness of which is not denied, but affirmed, and the reality of which to us is the very evidence of the want of life of MAN: their not being felt by us to be, as they are, the appearances of a different reality, showing man's defectiveness. It might be urged that no one leads so natural and common-sense a life, as he who best knows that he is living face to face with eternity and all spiritual things, and that a rectification of his own condition would make him feel himself to be so. It needs only a liberation from the chains forged by speculation and hypothesis, to make it most easy to us to recognise in all our consciousness a spiritual cause, and a deadness in ourselves. But assistance in overcoming the natural feeling, which makes this conviction difficult to acquire against our preconceptions, may be derived from the course of man's thought upon other subjects, and especially from the history of astronomy. Remembering that in the one case the intellectual apprehension alone is concerned, and in the other the actual being of man, the progress of astronomical discovery may serve to illustrate, in almost every detail, the course of man's learning that nature is spiritual, and that he wants life. We feel it absurd to be told that this is the spiritual world. According to all that we believe it is certainly not so. But we believe that the starry universe is infinite, or at least inconceivably vast in its extent in space; we reject with scorn the idea that it is confined within a petty sphere around the earth. Yet the wisest of men before Copernicus could not have believed the universe to be as we know it to be. It would have seemed as absurd to them to be told that the universe is infinite, as it is to us to be told that it is spiritual. And why? Simply because they ascribed to the starry heavens a condition which belonged not to it, but to themselves. On the score of their own feelings and perceptions, they believed the heavens were moving round the earth, and were forced therefore to conceive of them as they are not. Nothing could render it possible for men to think rightly of the universe in its relation to space, but the accepting for their own a condition which they perceived, and only could perceive, as existing in that universe. Just so it is with us. So long as we conceive a deadness in nature we cannot think of it as it truly is: but if we will accept that condition for our own, then there is no more difficulty. When men ceased to attribute their own motion to the universe, it expanded to the Infinite; if we cease to attribute our inertness to the universe, it rises to the Spiritual. Self-abnegation is the law of knowing. The universe cannot be infinite if it be revolving round the earth; it cannot be spiritual if it be inert. Is it a dead universe or a dead humanity; a revolving heavens or revolving earth?

Again: it may be asked, Are these things that we perceive by sense spiritual, or what are they, and why do we perceive them? To this it is to be replied, that the spiritual fact acting upon us, being such as we are, causes us to perceive in the way we do; but that the impressions we thus receive do not correspond with that which truly exists. Man's own condition has to be considered; it makes that which we feel to be different from that which IS. These objects of sense are the phenomena resulting from the relation of the fact to us, but not themselves the fact. They are the mode in which man perceives that which is. The specks of light which we see in the heavens are the appearances which result from the relation between the heavenly bodies and ourselves, but by no means do they correspond to those bodies. The heavens are not as they are to us. A very different thing acting upon us makes us perceive that appearance; compels us, while in ignorance, to believe the appearance to be the truth.

Our perception, our necessary belief in the world as physical, until we have learnt why it appears so, our being affected by physical things as we are, so moved by them and so deceived — that they, although not truly real, but only phenomenal, are real to us, and determine our entire experience and life; all this is part of that work in respect to man in which his relation to the spiritual world consists. The astronomical fact is not those little specks which answer to our perceptions, but that mighty universe which we have learnt from them. So the true and absolute fact of nature is not these physical things which answer to our perceptions, but that higher fact which has to be learnt from them. By man's littleness and deficiency, the impression nature produces upon him is below the truth of it. We have to remember this before we can think of it aright.

The problem presented by astronomy to man, and the mode of its solution, are an image of the larger and higher problem presented by the world, and of the mode in which its solution is effected. Our perception being modified by an unknown condition affecting ourselves, we have to learn what that condition is. There is only one way in which such a problem can be solved :—

The subjective element must be recognised as subjective, and transferred from that which is apart from ourselves, to that which implicates ourselves. In astronomy the history of the process is simple, and its essential features clearly marked. The appearance of motion due to man's own condition was observed, investigated, the relations which exist in respect to it accurately noted, under the conviction that the truth corresponded to the appearance. From this work of observation arose hypotheses, which were necessary to represent the appearances observed. These were the epicycles. The planetary motions were so irregular, owing to the combination of their motion around the sun with their apparent motion round the earth, that an immense number of revolving wheels were supposed, in various relations to each other, by the combined motions of which the apparent motions could be accurately conceived. For the epicycles were afterwards substituted, first a motion of the planets round the sun, and finally the twofold motion of the earth.

It is submitted that in respect to the inertness we perceive in nature, science as a whole has the same work to perform as astronomy performed in relation to the motion we perceive in the heavens: and it is here argued that this work is performed in the two cases in precisely the same way. The appearance is observed, investigated, and the relations which exist in respect to it accurately noted, under the supposition that it is the fact. Hence arise hypotheses even more numerous and complicated than those of the old astronomy: they are necessary to represent an inertness affecting ourselves as if it existed in the universe. These hypotheses, expressing the observed relations, constitute the substance of science as it exists at present. They accurately, and in the best way, represent that which is the phenomenon, that which man perceives. But so did the epicycles. Why, then, were the epicycles rejected? Partly for the reason that they became too complicated to be endured. They taxed the mind of man beyond the bounds of possibility, and the simpler conception took their place because it was simpler. So with science: the hypotheses with which it is encumbered are become too complicated. All these hypotheses are rendered necessary by the supposition that nature is inert, and the simpler conception that the inertness is man's, has claim to take their place because it is simpler. Our present science represents an astronomy that leaves the earth in the centre. As the wrongness of that conception was made manifest by the suppositions which it rendered necessary, so the wrongness of our conception of nature as being truly inert, instead of only being felt so by man, is manifested by the suppositions we are forced to form. We are compelled to admit our natural idea untenable. The fact cannot BE as it is to us.

We can maintain our natural conception of nature only so long as we have an unlimited indulgence in hypotheses, and frame a new supposition, of property, or principle, or law, for every fresh phenomenon that is discovered, as the ancients invented a new epicycle for every new irregularity observed in the planetary motion. When these suppositions are inquired into, and tested whether or not they can truly be, the case begins to appear different.*

---

* See 'A Speculation concerning Matter,' by PROFESSOR FARADAY: *Philos. Mag.* The celebrated Dr. Young expresses his dissatisfaction thus:—'It has been of late very customary to consider all the phenomena of nature as derived from the motions of the corpuscles of matter agitated by forces varying according to certain intricate laws which are supposed to be primary qualities, and for which it is a kind of sacrilege to

In astronomy men admitted so long that the motion was in the heavens, because it was tacitly assumed: attention was not directed to that question. When the inquiry was once distinctly raised, it could be decided only in one way. So have we admitted so long the inertness we certainly feel, to be in nature, only because it has been tacitly assumed to be so. The question has not been asked whether it truly is so or not. When once attention is fixed on it, and the inquiry distinctly raised, Is there *inaction* in nature, or inadequate apprehension on the part of man? can it be decided except in one way? Fairly to ask the question is the difficulty, not to answer it; to free ourselves sufficiently from conclusions which have always been taken for granted.

The argument which has been used respecting the Copernican Astronomy, that the senses do not deceive us in respect to the apparent motion, but give us an impression which is equally consistent with either of two explanations, applies in the same way to the question whether the inertness be in nature or in man. Nature need not be inert if man wants life; our perception of inertness cannot prove it to be in nature rather than affecting man.

---

attempt to assign any ulterior cause. . . . When a geometrician has translated this language into his own, and converted the formula into a curve with as many flexures and reflections as the labyrinth of Dædalus, he imagines that he has depicted to the senses the whole procedure of nature. Such methods may often be of temporary advantage as long as we are contented to consider them as classifications of phenomena only; but the grand scheme of the universe must surely, amidst all the stupendous diversity of parts, preserve a more dignified simplicity of plan and of principles than is compatible with these complicated suppositions.' — YOUNG'S *Lects.* Kelland's Ed., 1845, p. 476.

It proves only that there is defectiveness; there is a deadness either in nature or in man. If we will not allow it to be in man, then we must affirm it to be in nature, which we find so abhorrent; but if man wants life, then nature rises into joy. Then it must be living. Spiritual and physical, active and inert, are simply living and dead. But this is not all: The force of the demonstration of the motion of the earth, against the epicycles, consisted in chief part in this: that it showed why our perception must be such as it is. The old astronomy took the apparent motion of the heavens and said, We perceive it so because it is so, and these are the conceptions which we must form respecting it. The Copernican Astronomy takes a different ground. It says, These are the facts, and therefore our perception must be as it is; the appearance must be this. In this attitude towards our perception, it has an infallible security for prevalence. For the human mind demands in all cases to know why the appearance must be such as it is. It demands to see its perceptions necessary. This constitutes indeed the necessity of hypotheses; but hypotheses cannot maintain themselves, for they deceive the instinct instead of fulfilling it. They do not show our perceptions to be necessary, but merely assert something on their authority. They are like a vicious argument in a circle; there is an appearance of proof where there is truly none. To say, 'We perceive nature inert because it is so, and these are the conceptions we must form respecting it,' is not to show our perception necessary, it is to make a hypothesis.

But the transference of the inertness to man puts these things in the right relation. We understand our perception to be necessary, and see why the phenomenon must be such as it is. The fact being the absolute not-inert world, with a deadness affecting man, the perception should be, must be, that of an inert world (that is a physical world) even as it is. In this the mind can rest, its demands are satisfied. The hypotheses have served their purpose.

The direct proof that the inertness perceived as external is man's, corresponds also, in part, with that which supports the Copernican Astronomy; and especially in this, that to admit the inertness man's renders possible a satisfactory belief respecting the universe itself. Knowing that the heavens are not revolving as they seem, we can understand and enter into the relations of its parts; it appears before us a reasonable, consistent scheme of things; the entire conception so commends itself to our judgment that the evidence amounts to demonstration. Even so, knowing that nature is not dead as it seems, man and his relation to the world are presented to us in a way which we can partly enter into and understand, and which so assures itself to our judgment and our feelings, that we cannot doubt the appearance has received its interpretation. We can think justly of man, worthily of nature. The problem of the universe embarrasses the intellect, pains the heart, cramps and constrains the thoughts no more. It is a thing, in itself, and for ever, certain that the necessity of nature must be love. An inert necessity must have been felt, must have been supposed to exist, by a being in whom there is defect; but the necessity that can BE is love.

Another respect in which astronomy remarkably

illustrates the doctrine that the inertness felt in nature is in man, is furnished by the very difficulty of admitting it. The ground of this difficulty is, or seems to be, that it is against our consciousness. We have a conviction, so intuitive, so apparently insuperable, that man is not unspiritual, is not inert. On the contrary, this appears to be his distinguishing characteristic. If we cannot be sure of this, of what can we be sure? All our life, all our thoughts, are moulded to this persuasion. We base it on our consciousness. This is, however, virtually the same difficulty with which the Copernican Astronomy had to contend. We are certainly conscious, or seem to be conscious, that we are at rest in space, and that the earth is immovable beneath us. The earth was to the ancients distinguished from all the heavenly bodies by being alone steadfast, and that conviction was based upon the strongest evidence that consciousness can afford. Why should not our conviction that man is distinguished from all the rest of the creation that he perceives, by being spiritual while that is inert, be a similar error? Why should not his defect make him perceive a universal defectiveness, to which he feels himself the sole exception?

Even yet it is strange to us when we reflect, that we should be borne so rapidly through space and have no consciousness of it; but we admit it freely on the evidence that observation has supplied. And chiefly on the ground that the admission is necessary to enable us to understand the universe.

The same evidence may make us admit that this is not man's life. For it should be remembered that it is against our feeling and consciousness, not against an inference or belief only, that the Copernican Astronomy has made good its ground. The intensest natural convictions, the strongest persuasions of sense, inevitably yield to reason and evidence. That is a law of nature. If the world is not physical, men will as certainly believe it as that the heavens do not revolve.

Nor should there be much reluctance. For what is it but to put ourselves out of the centre, to be content to conceive of ourselves as subordinate and not chief, as being little parts of a greater whole instead of the end for which all exists?

When has it been found that humility, speaking in the name of reason and observation, has deceived us? We think too much of ourselves. This gives the fatal bias to our thoughts; is the judicial blindness of our eyes. God punishes us for pride by ignorance and error.

Let us remember that the aversion to admit the universe not revolving was of old as great as can possibly be ours to admit it not inert. No intensity of feeling, no apparent absurdity or impossibility in the idea, or firm conviction of the contrary, can lend any weight to the argument. Observation and the sound use of reason are the sole arbiters; our convictions and feelings and necessary persuasions are nothing. Rather, if they must be taken into the account, they are on the wrong side; for they are the fruits of ignorance, they are measurements of infinity by finitude.

We require to know why, if man be inert, our consciousness is such as it is. Why do we feel that our will is free? No theory has any claim to acceptance that does not account for this feeling. The question of Freewill will be discussed hereafter;*

* See Book III., Chap. vii.

here it may suffice to observe, in general, that it finds a perfect solution if the spirituality of the universe be granted. Man and nature cannot both be inert, but the inertness may be in man if it be not in nature. We think man free and nature not free. The consciousness of our own rest, and perception of motion in the heavens, affords a striking parallel. That man's will is free may be granted, if that form of expression be held to be of value. He has a relative freedom; hence his capacity for virtue and for criminality; but that this constitutes true freedom for man is entirely another proposition. It sounds strange with the words of the New Testament in our memory to hear the freedom of man affirmed as a Christian doctrine. 'Where the Spirit of the Lord is there is liberty.' Man is free just in so far as he has life.

In another respect astronomy aids our thought. The inertness that is affirmed of man is not such inertness as we seem to perceive in nature. That physical inertness does not, cannot, exist at all, it can only appear. It is a phenomenon due to man's inertness, but is not the same thing.

Even so the motions that appear in the heavens, real as they are to those who do not know the motion that causes them to appear, do not and cannot exist. They are appearances only. The truly existing motion is of another kind, of such a kind as to necessitate that appearance, but not the same as the appearance. Man's inertness is such as to cause a physical inertness to appear to him in nature, but it is not that physical inertness. The inertness of man is spiritual, real, actual; a true, absolute death, not a phenomenal one. Physical inertness is phenomenal only. From the true inertness come self-will, arbitrariness, and sin.

Yet by the study of the merely phenomenal motions in the heavens, which have no existence, but are only the impression produced on us by another motion of a different kind, the truth was discovered. From the study of effects comes the knowledge of causes. Even so from the study of the merely phenomenal inertness in nature, comes a knowledge of the true inertness which affects ourselves.

And the mode in which science effects this result is beautiful. For as astronomy dealt only with the motions of the heavenly bodies, having no possible regard to their essence; that is, chiefly with the subjective element, which is thereby discovered to be due to ourselves; so does science deal chiefly with inertia under the form of cause and effect. Science puts away the consideration of essence or being, and regards only causative connexion. It studies emphatically the subjective element, the inertia. Hence its adaptation to reveal to us its source in man. As the stars are at a distance from us, so that we can only observe their motions; so does science as it were put nature at a distance, and set aside all more intimate questions as to what it is, to study one particular condition. It deals with inertia alone; with causes and effects; phenomena and laws.

Yet this last expression is true only in a limited sense. It is true of the present, but not of that which must be. The true work of science is to discover facts. But the other position must be taken first, even as the epicycle, or apparent, astronomy must have preceded the Copernican. The old astronomy dealt with appearances and their laws alone, which it presented with the most truthful fidelity. To this day no better method can be found of representing those apparent motions. But the interpretation of those appearances reveals that which is, in relation to astronomy, the fact. Nor can science end in phenomena and laws; its destiny, its instincts, call it to a worthier work. As well might astronomy have left the earth for ever in the centre of the universe, and contented itself with the exhibition of appearances, and construction of theories which should account for them, as science leave nature inert to our belief, and end its work in manifesting phenomena and laws alone. Science abjures the inquiry after essences only to avoid false essences; first the fact must be made known before its essence can be inquired into. To seek the essence of a phenomenon were too great a mistake; it only appears to be. This is the abasement which comes before exaltation, the self-control and humbleness which are rewarded with unforeseen success. The astronomers of old little foreknew what work they were preparing for, what higher truer knowledge than ever they could conceive would flow from their labours. As little could the noble army of martyrs who have created science have foreseen the result their labours would achieve. For when the fact of nature is seen to be not inert but spiritual, then does science deal with the fact, and no more with phenomena alone.

In astronomy, again, we see that the false conception of the universe was overthrown by the observation of *relations*. For the relations which observation discovers belong rightly to the truth, and not to the appearance; and will not accord with the false conception. The attempt to harmonize them with that false conception necessitated the hypotheses which became at length so manifestly false.

So does science by the observation of relations in nature overthrow the false conception we have formed of it. For those relations belong rightly to the spiritual fact. The attempt to harmonize them with our natural impression necessitates our having recourse to hypotheses which we feel must be false; it produces a tension, a strain upon the mind, which ends of necessity in making us give up that false conception, natural, and at first unavoidable, as it is. Science teaches us that nature is not such as it appears to us; because if it be, we must believe things which cannot be believed, we must invent hypotheses which will not bear the test of examination, and at the complexity of which our natural instincts revolt. The oneness, the necessity, which science discovers in nature, belong to a fact that is not inert but truly active; they belong to holiness. It is by science, not by speculation, that the life of nature is made known, because it is by observation only that those relations become known by which false conceptions can be rectified.

Lastly: our astronomical knowledge enables us to understand how easily, when once we are familiar with the thought, we may practically and consciously recognise through all our experience a spiritual fact, in spite of our natural feeling. We have no difficulty in always thinking of the heavens as they are, and not as they appear. The apparent relations always suggest to our thoughts and feelings the true relations. We do, without any embarrassment or confusion, recognise what universe we live in; and are literally in a different universe from those who have not astronomical knowledge. Just so easy is it for any man to be consciously aware that the universe is spiritual; just so naturally may the apparent relations suggest to him the true ones, just so literally may he be in a different world from those who do not know that it is only in appearance physical. If we will remember that man wants life, as we remember that the earth is not at rest, we can perfectly well understand, and always be conscious, that the fact of nature is spiritual. All our instincts and native tendencies combine to enforce this belief; to the child and the utterly ignorant, the world is always spiritual, though in a false and perverted sense.

The work of science, in the discovery of invariableness or law, is not to exclude spirituality or action, but to give to it its true meaning of holiness, and teach us that the true spiritual is not that which man has, but that which he wants. Science proves nature different from ourselves, but places it not below us, as we think, but above. Man has to rise to become one with the FACT of nature. There IS not that inert existence which he feels to be.

Nothing is so repugnant, so impossible, as truly to believe the universe to be such as the theory of an external inertness represents it to be. It is manifestly more. Nature cannot be dead. We cannot help speaking of her life, inconsistent though it be. The difficulties with which science has so constantly to strive; the obstacles which theologians and poets so obstinately put in her way, are but the expressions of this feeling. Why do men so determinately maintain a special vital force, not identical with physical forces, but because they feel that life is truly spiritual, and will not have it made mechanical? Granted theirs is a blind and unwise struggle; that they deny the very spirituality they seek to maintain, and treat their best friend as an enemy. Not the less speaks humanity in them. Life is spiritual, and nature lives. Rather, far rather, will men admit man to be dead than the universe, when once they see that the question comes to that issue. For the point to be decided is not whether there be a deadness at all, within us or without. There is a deadness: we perceive it, and are conscious of it ever. We have embodied it in our language, asserted it in our philosophy, made it the corner-stone of our science in the doctrine of inertia. The deadness is the great fact of our present state of being, that which gives it its entire character. The assertion of a death is no new doctrine; it is no doctrine peculiar to religion. The only question is, where is it, in nature or in man? absolute or relative, affecting the universal work of God, or our miserable selves? Where is the WANT, the necessity for being altered? Is nature wrapped in darkness, or is man blind? This is the simple choice we have to make. A recognition that we are in the spiritual world does not demand of us so great a change in our conceptions as has been already accomplished by astronomy.

Nor does our understanding that the phenomenon is not the fact make any difference to the phenomenon itself. Our impressions are not altered; the only question is concerning the interpretation we put upon them. We perceive the universe as inert. Why? Because it is inert, or because our impression does not correspond to the truth by a defect of man's own being? This is almost too simple to lay stress upon, yet there is apt to be a misapprehension respecting it. The sun rises and sets to us as it did to the first of men. If it did not, we could not affirm the revolution of the earth. If nature were not inert to us, we could not affirm the deadness of man. The appearance is not altered by our better knowledge: the phenomenon is not made less by our knowing the fact, but more. It is shorn of no glory or value that it possessed, but receives an added lustre, a new significance. To know that the fact of nature is spiritual leaves us all that is in nature, but adds to it infinitely more. We do not thereby escape from the state which makes it physical to us, but we are freed from an illusion. The spiritual world must and should affect us as it does. To be affected otherwise, either man must be different, or the world must not be spiritual.

Man's defect is not in his perceiving the world as physical, but in his perceiving it as a reality; in his not feeling it to be phenomenal only: even as our ignorance is not the cause of our perceiving the heavens move, but of our thinking such motion to be real. From this state we cannot escape by any action of our own, nor is it desirable we should escape: but we can recognise the truth. We can think more rightly, though our impressions remain the same. So we are every way advantaged, and, especially, better prepared for action.

From our false feeling we learn what man's state is. We are such that the spiritual is physical to us, the active inert, the living dead; love a mechanical necessity. Such is man; such his defect; such his necessity for being made new. Here is the secret of his pride. Because he is dead, he sets himself up as the centre of all things, and feels himself exalted as such a king. He admires himself, extols himself, seeks to subordinate all things to himself, must make all things contribute to his pleasure; he must get all he can, must exercise his arbitrary will, will yield nothing, nor forego, nor sacrifice: all of which is the opposite to God.

He does not know that all this is from a miserable want: that as, through our own motion, the heavens revolve about the earth, and each man feels himself the centre of the universal sphere, so the secret of self-exaltation, self-will, self-regard, and self-assertion, is inertness. He says I am free, and nature is my slave: he does not know that this is death. Should he not rather say: In becoming one with that which nature IS, I live?

CHAPTER IV

OF KNOWING.

The inseparable propriety of time, which is ever more and more to disclose truth.—*Adv. of Learning.*

THUS astronomy exhibits an instance of a false belief respecting the universe, due to man's own condition. A belief established by universal consent, fortified by powerful arguments, and lasting many ages. Yet a belief fertile in practical evils, and necessary to be removed before man could use the world aright. Astronomy shows us also the simple and natural mode, of observation and learning from nature, by which such false beliefs are rectified. It should not therefore be urged against the opinion that the fact of nature is spiritual, that there is universal belief against it, and a natural persuasion of the strongest kind; nor should it prejudice the inquiry that so long a period has elapsed without the error being rectified. All these things we know may be; they have been before; it is natural that they should be. And if more ages have passed before man learns that he wants life, than before he discovered that the earth was not steadfast in the centre of the sphere, it may be remembered that the work is greater, and demands a larger preparation.

And if there appear to be strong arguments against this opinion, and much difficulty in admitting it, it may not be amiss to recall to mind that the true astronomy, basing itself upon the one certain argument that the perceived motion could not be in the heavens, yet met with many difficulties, and was opposed by strong arguments. Nothing could persuade a man so admirable in all respects, and so well qualified to form a right opinion, as the Astronomer Tycho Brahe, that the earth was not at rest. The firm persuasion of our steadfastness, except when we move or are moved relatively to the things around us, he could not give up; as hard that was to him, as it seems to us to give up the persuasion of man's life, except when he physically dies. Apparently he could not entertain the conception that man might be either at rest or moving to these things, and yet be not at rest, truly and in the strict meaning of the term: even as we find it strange to think that man may be living or dead to this earthly life, and yet not truly living in the strict sense of the word.

More striking still, Bacon himself, the great inaugurator of physical discovery, who led the van in man's deliverance from the persuasion of his own knowledge, rejected the doctrine of the earth's motion; not lightly, nor from mere prejudice, but on mistaken arguments drawn apparently from nature. He says: 'So we may see that the opinion of Copernicus touching the rotation of the earth, which astronomy itself cannot correct, because it is not repugnant to any of the phenomena, yet natural philosophy may correct.' It seemed to him that there were proofs, from other grounds, against it. Even so it might appear to us that, though an inertness in man instead of in nature could not be disproved from science, because it necessarily agrees with the phenomena, that is, with an absolute conformity to law and apparent passiveness in nature, yet it might be set aside by arguments drawn from other sources. Bacon's example, therefore, may teach us caution. The motion of the earth, proved by astronomy, refuses to be disproved on any other ground. May not a deadness in man, based on the simple argument that the perceived inertness cannot be in nature, compel a like assent?

Doubtless there were many things the Copernican Astronomy could not explain. Doubtless it was not, at first, fully reconcilable with all that was justly held: but also, demands were made upon it that had no claim to be regarded, and arguments were used against it, founded on opinions which further examination overthrew. The experience of the past may teach us patience. Can it be that our impressions of the universe should need no correcting; is it not certain they demand to be elevated and enlarged? Is it evidence in favour of our notions, or not rather truly against them, that man, such as he has proved himself to be, has been obliged to entertain them?*

---

* 'By prejudices of opinion,' says Sir John Herschel, 'we mean opinions hastily taken up, either from the assertion of others, from our own superficial views, or from vulgar observation; and which from being constantly admitted without dispute, have obtained the strong hold of habit upon our minds. Such were the opinions once maintained that the earth is the greatest body in the universe, and placed immovable in its centre, and all the rest of the universe created for its sole use; that it is the nature of fire and of sounds to ascend, that the moonlight is cold, that dews fall from the air, &c.' [May we not add, that nature is mere dead matter?] 'To combat and destroy such prejudices we may proceed in two ways, either by demonstrating the falsehood of the facts alleged in their support, or by showing how the appearances which seem to countenance them, are more satisfactorily accounted for without their admission. But it is unfortunately the nature of prejudices

But with regard to all the analogies which may be urged from our false impressions respecting objects of sense, to support the conclusion that nature in its true essence is not such as we feel it to be, it must be remembered that they are but analogies. They are used only to render the bearing of the argument more evident, and to show how in human life, as elsewhere in nature, one thing prefigures and prepares for another.

The cases are on a different level. In the one, an accurate conception is substituted for a conception conformed to a merely sensuous appearance; in the other, a belief in that which is not to be conceived is substituted for a belief in the existence of that which is conformed to our mode of thought. Which indeed is but to say that the very essence of being is above our power of conceiving.

The position affirmed is that the fact of nature does not correspond to our conceptions, even as we know it does not correspond to our sensuous impressions: and that in respect to our conceptions, as to our senses, this truer knowledge is acquired through the examination, and testing by observation and reflection, of that which we have believed to be. When it is proved that things which are to our sense cannot be that which truly exists, we abandon

of opinion to remain after all ground for their reasonable entertainment is destroyed. Against such a disposition the student of science must contend with all his power. Not that we are so unreasonable as to demand of him an instant and peremptory dismission of all his former opinions and judgments: all we require is, that he will hold them without bigotry, retain till he shall see reason to question them, and be ready to resign them when fairly proved untenable, and to doubt them when the weight of probability is shown to lie against them. If he refuse this, he is incapable of science.'

the belief without hesitation, and infer a different existence. We regard those impressions no more as authoritative, but look upon them as portions of a system of things to which they belong; and in reference to which, though not answering to the truth, they are such as we ought to have, such that we can from them, by due consideration, infer what the truth must be.

So should we act when it is proved that the things which present themselves to our conception cannot be the fact of nature. We should have no hesitation in giving them up also; and in inferring an existence, and a condition of our own, which should cause us rightly to have such conceptions, although not themselves answering to the truth. Our necessary conceptions of nature are such that we may, by due consideration, infer the truth from them. The fact which makes us conceive, as we do, a physical world we can certainly infer, by taking into account man's own condition, to have quite different properties. Our conceptions should be placed on the same level as our sensuous impressions, not as themselves authoritative, but as supplying the means of trustworthy knowledge. Thus, if the moon be not truly a bright disc, although that is what we must see; so neither can the universe be truly an inert existence, obeying passive laws, although that is what we must conceive. If we cannot think of it otherwise, that is of little moment; we may know it none the less to BE otherwise, and by the necessity of our thought know also something respecting man. Even so we cannot see the moon otherwise than as a disc; but we know it to be otherwise, and by the necessity of our sight we know also something of ourselves.

The necessity of our thought is like the necessity

of our sensuous perceptions, not an authority, not a thing to which our belief must be conformed, but a fact of our experience on which our true knowledge must be based. Both alike bespeak a cause, an existence, but are no evidence of what kind it is.

The fact of nature cannot be thought, cannot be presented to the intellect. But this is not strange: it must be so. That fact is not, therefore, less the reality to us, nor is it the less needful for us to remember of what kind it is. It is admitted that that which is true to thought cannot be presented to sense, but we know it is not the less to be regarded on that account. So that which IS (the true and absolute fact of nature) is none the less to be regarded because it cannot be presented to thought. It cannot be within the scope of thought: indeed it is more to us, and more real, and more truly known on that account. Being, as apprehended by our intellect, should be a deadness driven by inert necessity. Thereby we know it and ourselves.*

And with regard to the existence of the things we touch and use, what is true of them is only that they truly have a relative existence. They are relatively to us: they have the same existence as our bodies. They are to man's feeling, to his consciousness. The existence of phenomenal things is not denied in any such sense as to leave a blank or vacancy,

* That which is thought, is not therefore known. Those things which we most truly know we cannot think. Can we think ourselves, our affections, our life, our Being? Absolute being cannot be thought. That which is thought is, of necessity, an idea, not an existence. Ought we to be able to put the universe into our minds?

as if it were space left empty. They exist as parts of the whole; but that whole is different from what we conceive. It is asserted that our apprehension of the universe is inadequate.

It should, however, be remembered, with reference to the illusions of the senses which thought corrects, that we have no other kind of evidence for those things which we find most real and certain, than we have for those which are proved to be illusions. The evidence we have of the earth on which we stand is not truly different from that which we have of a bright disc in the heavens; it is more, but it is still the evidence of sense. We see the moon as a disc, we see and touch the earth as rock, or water, or tree, or house; but touch may be deceptive as well as sight. We have but two united senses; we have not a security against the deceptiveness of sense. We know only that the earth is to us. It supports us, answers to our efforts, to our conceptions. And if there be in us an extreme assurance, that that which sight and touch unitedly affirm must be truly that which IS, we should remember that there is in us the same natural assurance, until it is corrected, of the certainty of sight alone. If men have been willing to determine, upon grounds of sound reason, what is truly indicated by that which they only see, surely it may be expected that they will be willing to use the like consideration with respect to that which they both see and touch. Truth does not come from clinging to our natural convictions.

And our finding these sensuous things always the same, or changing according to definite laws, so that we can calculate upon them, use them, and feel them reliable, firm, and true: this does not affect the essential nature of that which IS. The fact, being ever the same, is ever the same to us; that which is to sense will be to sense again, that which is to thought will be the same to thought again. There is that which causes us to see a bright disc in the heavens, and so long as it is the same and we are the same, so long shall we see that disc; but that which we believe in, through such seeing, is very unlike the disc. We know that if we saw it truly, we should find it extremely different. For our perception of such a disc there must be more than we perceive.

Never, indeed, can we account for our perception by the existence of things which correspond to it. If the fact were not more than that, never could we perceive so much. Nay, we should certainly not perceive at all. How should inert things make us perceive? The fact of our perceiving as we do proves the existence of more than is perceived. Can our apprehension be equal to nature's excellence? Do we not, even in respect to every single thing, have to believe more in that thing than we immediately perceive? How then should it not be the same in respect to the sum of all?

For our experience to be such as it is, there must be, to our feeling, inert or physical things. We must perceive them and act upon them; they must be the realities of our existence. How should this be? Inertness is opposed to being: there cannot truly *be* inert existence. We must, therefore, perceive as inert that which is not inert; only so can the things we have to do with be passive, dead, material, such that we can exercise force upon them, and find resistance. Only in one way can we have perception and experience of inert realities: there must BE, apart from man true not-inert existence, and inertness in him: then to him there will be inert existence. The existence without, the inertness within. Is not this the solution of our perception? Inert things thus will be to us; a physical world, answering in every way to our feeling and our action. And not only so; but thus alone can it be, that the things which are the realities to us should cease and pass away. For our activity, progress, enjoyment, for this life of man, the things that are to us must change; they must have been, but be no more. They must not BE, but be temporal and fleeting: the forms under which an unchanging existence is perceived. With true eternal being around him, and defect in man, inert and passing things are his realities. He dwells amid phenomena, and lives a temporal and earthly life.

Do we ask: How should man be in an inert world? Let us ask: How should he be in a revolving universe? These two questions admit of one reply. He is not so. The universe cannot be revolving. Let the universe, therefore, stand fast, and man revolve. So shall be to him day and night, rising and setting suns, noonday brightness for his work, and solemn revelations of the stars to lead him up to God.—The universe cannot be dead. Let the universe be living, therefore, and man be dead. So to him there shall be a world of passive laws and lifeless uniformity, a world subject to his control, invitant to his energy, full of deep lessons to his heart.

CHAPTER V.

OF BEING.

God is Love.

WE have necessarily inferred, from our experience, the existence of an inert world; conceiving that the fact corresponds to our impressions. But what we are compelled to infer depends, in every case, upon our knowledge: only when that is complete and exact, can an inference, however necessary, or belief, however unavoidable, possess correctness. A person ignorant of any essential circumstance, in any case, necessarily infers an erroneous conclusion. The necessity of a belief has no necessary relation to its truth; true and false beliefs are equally necessary to instructed and uninstructed persons respectively. Nor is there any conscious difference to the mind, in respect to its necessity, between a true and a false belief. They can be distinguished only by being tested. A true inference proves itself true on examination; a false inference is found by the same means to be false, and proves thereby ignorance on the part of him to whom it has been necessary.

These are very obvious considerations to apply to our opinions respecting the world; nor would there have been any difficulty in applying them, but for one circumstance, which has seemed to distinguish those opinions from our opinions on all other subjects. It has been thought that the belief respecting the world, which we derive from our consciousness, must be held infallible, because, if it is not so, the sole basis for certainty is taken away. It has been supposed that if such belief is untrue, we not only are under illusion, but are hopelessly and inevitably so: under an illusion from which man can never escape. It is not denied, on the one hand, by any man who has considered the question, that our consciousness might be caused in a manner quite different from that which we necessarily suppose; yet it is maintained, on the other, that it cannot be so, or else man would lie under an irremediable delusion. The argument is, that we cannot believe that his Creator would have made him so; would have given him capacities, instincts, and desires only to mock and to deceive him. And the argument is in itself a good one. We are persuaded that He has given to man the means of knowing the truth, a basis for certainty; and cannot have left him hopelessly under illusion. If man truly had no means of correcting his first necessary impressions respecting the nature of the world, there would be a fair basis for maintaining, on the ground of the divine character, that those impressions correspond to the truth.

But this position, though it may be well understood, and the grounds of it appreciated, may be seen to rest on an imperfect knowledge. It is not to be opposed, for it merges itself into, and becomes, a different one, when the work of science is rightly apprehended. It is not a true assumption that man has no means of correcting his necessary impressions respecting the world. If they are false, he is not left hopelessly under illusion; he is not mocked by his Creator. The means whereby that illusion is escaped from, as in the case of all other illusions, are at his command, in investigation, observation, and inquiry; in the right use of his natural powers. In reasoning upon this subject men have overlooked, as they naturally must have done, the true bearing of science: they have conceived it wrongly; placing it in subordination to their natural impressions, instead of recognizing in it a power to correct them. The adaptation of science to this end escaped their attention, and finding no other means by which our impressions of nature could be corrected if they were wrong, the infallibility of those impressions became an unavoidable inference.

Science operates to correct our natural impressions of the world, in the same way as all erroneous natural impressions are corrected; by increasing our knowledge, by causing us to see more truly the relations of things, by proving to us that our conception will not answer to the facts, but leads us into difficulties from which an alteration of our conception delivers us. Science proves nature spiritual and man wanting in his true life, just as a child learns that a reflection of himself in a glass is only a reflection. There is nothing peculiar in the process, it is only on a larger scale; that is, on a larger scale relatively to us.

It will be a happy day for man when he clearly understands that that which is great and important to him is not therefore truly great, or different from that which he calls trivial; for well we know that there is nothing which so keeps man back from knowledge as his pride, nor is that pride ever more fruitful of mischief than when it clothes itself in the garb of humility. But how can it be other than a false humility, that presumes to fix boundaries to the possibilities of human knowledge, and says to God, Hitherto shalt thou go in instructing humanity, but no farther. Upon what basis can such a position rest, but on the assumption that we cannot be deceived, and that the fact must be as it appears to us; that what we cannot see any means of doing never can be done? Of all forms of self-assertion, none is more arrogant, hardly any is more thinly disguised. For all this means that we cannot admit it possible that we are ignorant and mistaken, unable to expand our thoughts to the true meaning of that which is around us:—we, assuming that we regard the world in the very best and truest way possible to man, and finding that our thoughts end in mystery, lay it down as certain that no man ever will regard the world more truly, and so escape the mystery. We are content to let a darkness rest upon God's world until the end of time, but not to admit that men hereafter may be wiser than ourselves. Nor is the inconsistency less than the presumption. For the denial that man can know, rests upon the assumption that he does know, that his mode of conception is correct.

Yet there has been a certain justification for this mode of thinking. For it is true that all attempts to explain the essential nature of things have failed. Thinking as we do, the essence of being seems hidden from our eyes in an impenetrable mystery; our faculties seem unfitted to grapple with the question. It was natural that men should say, we cannot know it. Natural but not right. The legitimate inference would rather have been: we must alter our way of thinking; we are conceiving the case wrongly. For we have fallen into a strange idea about mystery. We seem to think that a thing can be mysterious in itself, and apart from mistaken or inadequate conceptions respecting it; forgetting that all things must be clear when they are known, and simple when they are understood.*

Mysteries mean errors: they arise from ignorance, and ignorance implies a false conception. To all but Omniscience, indeed, there must be mystery; but the meaning of mystery is, none the less, that we are thinking wrongly. If we knew more, the mystery would be gone, for we should conceive differently. Therefore mystery always reveals to us wrong thoughts. The essence of things is mysterious to us because we are not thinking rightly respecting them. We should think differently if we had more knowledge. But science is adapted to remove the mystery from nature because it adds to our knowledge, and so helps us to think differently. Explanations and philosophical speculations necessarily fail in their attempts, because they do not add to our knowledge, and cannot therefore alter the false conception from which the mystery arises; they make the mystery manifest, but cannot remove it. Because philosophical speculation fails to diminish the mystery of the world, it has been laid down that the mystery cannot be diminished. But this is too hasty a conclusion. There was no adaptation in the instrument.† Explanations naturally fail; it is by their failure that men are driven to investigate and learn.

But science holds a different relation, and places

---

* Forgetting also the scriptural use of the word mystery as denoting a thing unknown, and not a difficulty.

† Well did Bacon say: *Equidem organum præbui.* 'I have furnished the instrument.'

man in another attitude. Science is an investigation of nature, not in its parts only but as a whole, and thus gives man the knowledge by which he may escape from the false conception which his ignorance has imposed upon him.

For it is the conception of EXISTENCE as physical, or inert, which involves in mystery the problem of Being. Of physical existence the problem never can be solved; all attempts must land us in deeper darkness, must make the contradictoriness more manifest. We are trying to think of that as 'being' which cannot be, but can only 'appear.' We are putting the phenomenon for the fact. All our conceptions, all our attempts to think, are baffled and brought to nought by this error; no hypotheses will fill the chasm, no imaginings hide from ourselves the consciousness that the very fact and essence of all things escapes us. Conceiving an inertness in the universe, a negation not relative but absolute, we are amazed that we cannot conceive what that BEING can be, to which inertness belongs. But why should we be amazed? How can inertness belong to BEING? Inertness is deadness. Here, in ourselves, is the Being to which inertness belongs; we know it but too well: that being which is the slave of passion, which obeys impulses, which does as it likes, not doing what it ought to do, and doing that which ought not to be done; which hears in its heart a voice saying, Thou art evil, and evil things await thee, for evil must be where thou art: that is the being to which life must be given that it may be inert no more; for which the Saviour must pour forth His blood, that with love, life may come.

The difficulty arises from man's false supposition of himself; he seeks to know by sense and intellect, which deal only with phenomena. Hence he fails and must fail: and says rightly, I must be different before I can know. But still his thought is wrong: true being cannot be thought: to know it is to BE. To know God is not to have an opinion, it is life. True knowing relates to being, not to thinking.* Man cannot think that which IS; but he can KNOW it, for it may be in him.

If the apparent inertness of nature be due to man's deadness, the course of man's thought must have been such as it has been. Experience gives sufficient evidence to justify this conclusion; for man does perceive according to his own condition, and he does, by observation, learn to distinguish the truth from his false perception. Just such a course as his has been must have resulted from a perception of the spiritual as not-spiritual, by defect in himself. He must have been firm in his native error; he must have constructed a science on that basis; he must have inferred an inertness without him; he must have erected hypotheses to account for the phenomena accordingly; he must have become embarrassed by the complexity and incongruity which they involved; he must have gone deeper into mystery with every step; he must have given up the problem in despair. All this has been and is. The conception that the world is spiritual, and only physical by man's defect, reveals the life of science.

Still more, it shows why men must have affirmed another and a higher world than this; must have

---

* To know, is 'to be one with.' The word has a meaning which the intellect cannot fathom.

affirmed this a degraded state and a preparation for another; why that doctrine has been so opposed and yet maintains its ground. 'There is a spiritual world.' 'There is no other world but this.' Both are true. There is no contradiction here; only the self-evident mistake of thinking that the world is such as we feel it to be.

There is, indeed, a scheme of things erected on the supposition that inertness exists apart from man. Phenomena have been set up as the reality, and all the suppositions required by that view duly inferred and asserted. The system is complete, and is freely on every man's tongue; it is the theory of a *material substratum.*

But it is not believed. Denied on all hands, and on the most various grounds, that theory appears to exist on sufferance, and to remain only because there was nothing that could take its place. For all those who affirm that our knowledge is only relative, deny it absolutely; those who admit that the essence of being is beyond our thought, deny it by implication; the mass of men have no thought of it; metaphysicians allow that it can be disproved; preachers affirm it to be an empty show, an unsubstantial dream. But its worst enemy is science. All the others, while they deny its claims, and sap its foundations, still leave it in possession. For that theory cannot be overthrown until its place can be sufficiently filled. We cannot give up believing the appearance, until we understand why it should appear. This demand science fulfils; revealing holiness through uniformity; love in necessity; life where we have conceived death, and death where we have fondly imagined life.

For by its very nature, and in all its tendencies,

science implies, and prepares us to recognise, the spirituality of nature.

Science sets aside and denies the authority of man's conceptions; renders him familiar with the thought that the universe infinitely exceeds in glory and majesty all that he could have supposed. Science accustoms men to admit, in that which they perceive, the presence of a fact entirely different, as when it teaches them to recognise in motion the cause of their sensation of music or of light; making them understand that nature is altered to them by their own condition. Science teaches men that they do naturally attribute to nature a defect which is their own; their own arbitrariness and variableness, for which observation substitutes a law fulfilled.

Especially does science teach that that which is in time IS not, but is only form. For forms only change. The fact of nature cannot be in time, for if that which IS may cease to be, science has no certainty.

Science deals with *action;* it recognises operation only, and knows nothing of inert substance; the doctrine of inertia, to which it is forced to have recourse, is abhorrent to its nature. Science confirms that voice which says within our breasts that this world which is so real to us is but a show; proving that all phenomena point to a higher fact which is not in them.

By science man is cured of his false notions of the spiritual. He thinks that he can conceive it by his intellect; he is taught that he cannot. He learns his own defect, that the true spiritual is not in him.

Science is prayer and answer. Man cries to God: What doest Thou, O God, that heaven glows with innumerable orbs, and earth's palpitating bosom bursts into ceaseless life? What doest Thou? And from the Infinite Heart the still small voice replies: I love.

Since there is in science a means by which man's natural convictions in respect to the world may be rectified, there remains no more reason for refusing to admit them to be erroneous. And there is a great relief to the mind in being able to take this ground. Man's life is brought into greater harmony and consistency with itself. For it is the law of our present state that we should learn truth through illusion. Nor can we, indeed, conceive it to be otherwise, without an entire alteration of our mental constitution. Starting from ignorance, error must have precedence of truth. Our whole state is one of illusion till we are delivered from it. If this appear strange to us and unnatural, it is proof that we think wrongly of the state of man. Illusion is natural if this be not man's true life; if for true manhood he have to be made different. In fact, man is under illusions which include all his being. For does he not find pleasure in that which may be harmful?—are not poisons sometimes pleasant?—are not most enticing enjoyments often disastrous and evil? But to have pleasure in a thing is to feel it good. Man, therefore, may feel that to be good which is not good, that to be evil which is not evil. His feeling of good is no proof of goodness. He is under illusion as to good.

And does not man often necessarily think that to be true which is not true? Does not his ignorance determine his opinions? That which he must think true may, or may not, be true. He is under illusion as to truth.

And again. Does not man of necessity think

that to be, which appears to him, although it may not be? He may feel that to be which is not.

But if man is by his nature under illusion as to good, and truth, and being, how should this be his LIFE?

There are two possible views which may be taken of the universe, conformably with the appearance: two interpretations which may be put upon that which we perceive and are conscious of. We may think, as we have been accustomed to think, that nature is a dead inert entity, subject to mere passive law, with one being of spiritual capacities and endowments, and he mysteriously failing, sinning, evil, falling short of all that he should be. The one being, worthy to be called a Being, marred, and lost, and evil. No true life except in man, and in him so strangely spoilt.

Or we may think nature perfect in spiritual life. An universe full of being that is true being, with no flaw, with no defect; but in respect to man this being wanting. Man the one defective thing. Not that the universe is imperfect by his defect, marred by his failure. That is part of its life. Only in respect to him is there defect. Only relative, not absolute. He is what he is, because life is to be given to him; his consciousness, his work, his action, have reference to a life that is to be bestowed. Viewed in relation to man there is defect. But man's defect must be—must be for this human love, for this human life. Without this inert consciousness self-sacrifice could not be, and in self-sacrifice is creature life. Man is to have true being; his deadness is made conscious, as it were, to himself, that he may be delivered from it. Therefore he feels it to be in all that he perceives—therefore the spiritual world is a dead world to him, the universe is so mean, and he so lofty.

According to our conceptions, there is a rightness in nature, but that rightness has no worth; man has worth, but he is wrong. Surely we are right in feeling this to be a dark and painful mystery. But where have we learnt that it is true? What evidence, what ground, what right have we to assume it? That is the *phenomenon*, that is what is felt by us. If it be so painful a mystery, why believe it, when it rests only on the assurance that we cannot be mistaken, and has no evidence but that we feel it so? For if we be wrong, we must feel wrongly; if even we be only ignorant, we must think wrongly. We do not rely on such reasons, on such evidence, respecting the simplest and most ordinary circumstance. Our feelings are of no weight unless we know and consider also our own condition, and our relation to that which is their cause. Why should we act against all experience and all reason, and assume that our impressions are correct while we are ignorant, that we can know without the means of knowing? Why explain instead of investigating, when our explanation fills the world with gloom? The evidence on which we take for granted that the universe IS such as we think it would not avail to establish the very slightest fact in our daily life: viz., that it seems so to us, without our having learnt or, inquired, whether there were any circumstances affecting the mode in which it seems to us.

That man should be under illusion only shows that there is defect in him. It is but the necessary consequence of a fact well known. In recognising that we have been under illusion, we do but assent to an admitted principle, which we might well marvel we had not recognised before. How should we, who without investigation cannot know one single detail of the course of nature, know without investigation the essence of the whole? How should we, who are deceived and under illusions constantly in respect to matters of the most ordinary import, know that we could not have been deceived in respect to the highest and profoundest of all? Is it not arrogance, the very extremity of pride? Can we wonder that, asserting confidently their own impressions, men wander in labyrinths, and cannot right themselves?

Man may take for himself in God's universe a lofty or an humble place: the one living being or the one wanting life. He may, in his thoughts, exalt God or exalt himself. In either case his natural impression, his perception, must be as it is; in either case he must seem to himself the one living being in a world of death.

It is a simple question, it might seem an idle and merely speculative one; Is the perceived inertness nature's or man's? But what practical issues it has, what a determining power. On the answer to that question depends the entire attitude of human life. Men wait to be delivered from illusion: they wait to know what the FACT is with which they have to do.

BOOK II.

OF PHILOSOPHY.

The improvement which remains to be effected in the methods of philosophizing can only consist in performing more systematically and accurately operations with which, at least in their elementary form, the human intellect is already familiar.—J. S. MILL: *System of Logic.*

CHAPTER I.

OF MAN.

Laudable faith consists in resolving to receive and acknowledge whatever there is good ground for believing, however contrary it may be to our expectations, wishes, and prejudices . . . . in listening to reason notwithstanding all the strange circumstances that tend to bias the mind the other way.—ARCHBISHOP WHATELY.*

IT might be thought that the idea of a deadness in man, affecting his condition in all respects, and causing his impressions and natural convictions to differ from the truth, would present difficulties to the understanding, and run counter to the feelings. But the case is not so. The contrary idea, that our natural impressions must be taken as true, subjects us to embarrassment and constraint. Clinging to these, we forge chains for our own hands; but to understand that the world is truly different from that which we feel it to be, more than it is to us, sets us free. Surely our apprehension of the universe must be inadequate; that which we think, no more equal to its truth, than that which our senses represent to us. We know that the universe is more than corresponds to our conception; what, therefore, can be more natural than that we should distinguish in our thought between that which truly exists and that which we can conceive? It can be no hard task to recognise in a new bearing the familiar truth that our condi-

* *On a Future State.* Sixth Ed. pp. 326, 329.

tion and relation to things determine the mode in which we are affected by them. We only apply to the whole that mode of judging which we have already applied to particular things. We do not consider that we know even the size or shape of any object until we have considered how far we are distant from it; we decide on the nature of nothing that we see or feel without reflecting what relations we bear to it, what our own condition is. The proposition, therefore, that the universe is not truly physical in its own being, but is rendered so to us by man's condition, involves no new mode of thought or principle of judging. It is not a speculation, but a question of ordinary evidence, appealing to the rules of judgment which are daily applied by all men, and used more reflectingly, and on a larger scale, by men practically engaged in scientific work.

For, when inquiry is made respecting the world, the primary answer is the same, whether it be held that the appearance does or does not correspond to the reality: the world exists. Then there arises a second question: Has that which exists those qualities which we naturally suppose, or are its true qualities other than its apparent ones? Is it, or is it not, necessary that we should take into consideration ourselves and our relations, to enable us rightly to appreciate what it is? Which question, indeed, is simply whether we should employ the means found necessary for arriving at true opinions in every other case, or should adopt another method, against which, in every other case, experience testifies. Shall we act according to experience and reason, or on some supposition to which these general guides of our conduct lend no sanction?

For in asserting that man's own condition determines the mode in which he is impressed by that which is apart from him, and necessitates his thinking it to be other than it is, until he has examined, and ascertained what the entire circumstances are, we do but recall to our thoughts an evident and admitted principle. All men, in theory, are willing to concede that our impressions of nature do not correspond strictly, or exactly, to the fact. The least consideration, indeed, suffices to make this evident. For the relations between ourselves and the fact of nature, on which the impressions it must produce on us depend, are not known to us beforehand, and the law of our perception, therefore, demands that our impressions shall be corrected by the discovery of those relations. There is no peculiarity or exceptional character in the case. Surely it is as natural that a world not physical should be physical to our perception by virtue of our relation to it, as that a world not at rest should be at rest to our perception. In judging of the size of anything, or of its condition as to space, we have regard to our distance, or relation to it in respect to space; in judging of nature in respect to its mode of existence, we must have regard to our relation to it in respect to our mode of existence. According as we are, so will that which is without us be to us.

Therefore, in truth, our case is thus with respect to nature: either we cannot know it truly as it is at all, and must be hopelessly under illusion, or we must learn to know it by discovering our own condition in relation to it, and interpreting the appearance it presents in conformity therewith. The well-ascertained laws of mental operation do not permit any other conclusion.

It is necessary thus to insist at length upon this point, because, simple as it is, the whole question of the spirituality of nature is contained in it. If this principle can once be clearly seen; if it be felt that a true knowledge of what nature is must depend upon a recognition of our own relation to it, as in every other case right knowledge depends upon such recognition of our own relation; if the question can be brought out of the domain of darkness and assumption, and be treated on the principles called, in all other cases, those of common sense, the entire difficulty is overcome. For the necessity of regarding the apparent inaction in nature, or that which is wanting in it, as the result of man's condition, hardly needs to be insisted on when once it is recognised that a regard must be had to the condition of man, and that some part of the apparent quality, or mode of existence, of nature is due to it.

Further proofs of this position will present themselves in their due place, but in truth it is difficult to see what other proof can be so convincing as a simple statement of the alternatives. For the inertness that we must recognise in that which is conceived as physical (think of it how we will or in whatever light we may endeavour to place it), must retain the character of being an absence or negation. Call it by what name we may, we cannot escape from this: in nature as it is perceived by us there is something that we must admit to be of a negative character. Men have named it inertness, or absence of action; but if we object to this, and prefer to regard it in any other way, its essential character will not alter, it remains still an absence. Therefore we must either admit a negation as absolutely existing; must conceive an universal inertness or absence as in some way created; or else that nature is not such as we feel it to be. But is not the appearance of an universal absence, or defect, simply the way in which we learn that we do not perceive that which is truly existing? When we examine and reflect upon the facts of the external world, we are compelled to think of them as involving an absence of something: what can this mean, but that we do not perceive that which truly is? An absence or defect may well be in that which we perceive, if our perception do not correspond to the fact; but how can it be in the absolute fact itself? Can we conceive a more exact contradiction than that an universal negation exists? or that an absence has been created? To hold fast to our natural impression, and refuse to correct it by admitting that there is more in the fact of the universe than we have supposed, drives us into the most obviously impossible positions. The idea of a true instead of an apparent inertness in nature, or that the universe truly IS not active, or physical, proves itself impossible the moment it is looked into: only through taking for granted and not inquiring can we have rested in that opinion. Whatever the truth may be, that cannot be true. And then the other question follows: If nature be not truly physical, why is it physical to us? What condition of ours is it that makes the not-inert to be felt as inert? Evidently it is some condition that makes the existence around us to be less to us than it truly is. It is a non-perception on our part; an ignorance. That which truly is in nature is not to us. We introduce the negation, the absence; the negative element, be it what it may, is ours; we must search for it within. This at least may be held certain. And how simple it is! We knew that the truth of nature must be different from our conception of it, because that conception cannot be adequate; now we know one respect in which they must differ. Our conception is of an inactive nature, though nature cannot be inactive. Our conception is therefore inadequate in this respect. The admitted general principle receives a partial application.

And in this all is involved; it should not be necessary to add anything more. Let the conclusion be held fast for a moment and considered. Defect in man causes the universe to appear to him such as it does appear, to be to him defective. How simple is the statement—nay, how self-evident, commonplace, and trite. Nothing can be less new, less doubted. In his heart no man thinks otherwise; by the very necessities of language no man can speak otherwise. Yet how strong an illusion holds us. Against all this, which we know, and are so well assured of, we cannot help maintaining that the universe is such as it appears. It is a strange contradiction; our nature seems divided against itself. That which in theory we give up most readily, in practice we cling to as if for our very life. We say willingly, that which exists is not such as it appears; but we dare not say, that which appears is not that which exists. Why is it that we are so mocked, so bound? Is it a mere solecism, a contradiction in our nature, a mystery we shall never solve? By no means. There is no solecism, no contradiction, nothing but that which ought to be. If the case were not as it is, there could not be that defect on man's part, which is the secret of the whole. For what is the source of the embarrassment but this: that the things which appear are real to us, that which is not truly the fact is the fact to us, determines and controls our being. Our existence, as we are, is in that which is inert. Here is the contradiction. Our reason and our feeling are at strife. We know, when we reflect, that the things that appear cannot be the things that are; yet we feel them to be—they are to us. That is our defect. That constitutes our world physical; makes the phenomenon the reality.

Thus there must have arisen the perplexity that embarrasses us, and makes us say: 'It is impossible to understand these things, and therefore we should not inquire, but must fall back upon the infallibility of our own impressions.' We feel that these things which are the objects of sense certainly are, do truly and really exist, although their existence can be disproved: but so we should feel; so the defect of man's being reveals itself. And the intellectual difficulty with which men have struggled so long and so vainly, as to the existence of the external world, arises in the same way. We have not recognised the defect, in respect to man, which causes us to feel as we do. Nor, indeed, could we until the problem had been worked out. For the discovery of that defect, the knowledge of man's own condition, is the result achieved by the work that has been done in ignorance of it. Even as the knowledge of man's motion in space, which is the key to the heavens, is the result of the work which was done in ignorance of it.

In all this there has been nothing peculiar or unlike the rest of our experience. It can appear so to us, only so far as our thoughts have been merely speculations, and not based on rational inquiry. By means of the fact that conditions affecting ourselves modify our perception, those conditions are made known to us. No arrangement in nature is more beneficent, or better adapted to its end, than this. Our own condition (which it is in some sense the most important of all things for us to know, not only because of its immediate interest, but also because such knowledge is the basis of all right apprehension of other things) is revealed to us by means of its effect upon our perception. If that which properly belongs to us had not this effect on the appearance of that which is without us, if our own state were not thus made apparent to ourselves, and brought within the sphere of observation and inquiry, we could never know it, we should have no means of learning it. By study of that which is without us, we must learn what we ourselves are. The steadfast stars alone could reveal to man the restless circuit of his little globe. He sees them revolving round the earth that he may know his own motion and its cause. So the spiritual universe, not-spiritual to human apprehension, reveals the defectiveness of man. The spiritual is felt by him as physical, he feels the appearance as the fact, that he may know his own deadness, what and whence it is.*

Nor is there in this representation anything speculative or unpractical. Let the principle be tested by a few familiar instances in which its true bearing may be more easily appreciated. We are apt to say, we see and feel these things to be physical; we know them to be so: meaning that the very facts that truly exist are so. But do we not see the moon to be bright? Do we therefore know it to be so? On the contrary, we know it not to be so,

* It is true the stars are not absolutely steadfast. They are so, however, relatively, and for the purpose of the illustration.

and find no difficulty in apprehending the truth, because we recognise the laws of our perception. But of old, before astronomy was rightly understood, men who saw the moon bright could not have been made to understand or believe that it was not so. They would have said: We see it bright, we know it is so. In all cases of perception we must feel convinced that the fact corresponds with our impression until we know the conditions which cause our perception to be erroneous. A straight rod, partly immersed in water, looks bent, and so we should believe it, were we not able to correct our experience by reflection.

In judging of the being of the world we must take into consideration the state of man. This is no abstruse idea; it is at the farthest remove from being speculative. When we place our hands, first in very cold water, and then in water less cold, the latter feels warm. We should naturally say: I feel it warm, it is warm: but would it be an abstruse or speculative thing to reflect that we must think of the previous condition of our hands; and that the water was not warm, although we felt it so? Whatever it may be to take into account what man is when we would judge what nature is, it is at least not to be unpractical; it is not to deviate from the rules and maxims of ordinary life; it is not to obscure a plain question by subtleties.

Still less is it to lose or be deprived of anything. A feeling is apt to take possession of us, and one from which we cannot immediately escape, to the effect, that if the appearance of the universe be not such as the fact is, then there is less than there would otherwise be; as if some 'existence' would be set aside. A little reflection frees us from this embarrassment, which indeed is not peculiar to this case, but arises continually with the advance of knowledge. The change of an incorrect opinion for a true one always involves the loss of something that was connected with the former. An idolater in learning better to understand the Divine nature loses his gods of flesh and blood, his solid, substantial divinities; and finds it, at first, difficult to understand that he has truly incurred no loss. Ignorance necessitates suppositions which knowledge sets aside, but meanwhile those suppositions have gained a hold upon the thoughts as if they were realities, and the parting with them is felt as a deprivation. The slave accustomed to his bonds misses his shackles in his first days of freedom. What loss is it to give up the less for the greater, to loosen the grasp upon the transient to lay hold upon the eternal, to change that which appears to us for that which IS? How can our thinking differently of nature alter anything in it: how make it less? What can we lose by knowing better? And especially, how can it be a loss to feel that nature is more than we have thought, to understand that that which appears as a want in it is from our ignorance? of what can this deprive us?

What we miss, and feel to be taken away from us, as if it were a possession, is the necessity of making suppositions, of inferring certain things. We need not any more suppose what we have hitherto been compelled to suppose. The opinion which necessitated those suppositions being changed, they are no longer necessary. We have altered our view, and perceive that the facts demand a different interpretation. That is all.

Nature is more than we thought. And man is more also; simply our view is enlarged: the infinite wonder and majesty of God, and of His universe, are more worthily revealed to us. We thought we were more on a level with them than we are. We brought them down to ourselves; now we seek to rise to them. Does it make man less, does it not rather at once exalt humanity and fill us with humility, to understand that the life of man is not yet ours; that true manhood is more than we possess, more than we have thought? What loss is it to know that we are wanting: is it not infinite and blessed gain, the first condition of all betterness? What loss to know that God will not, cannot, leave us as we are, but will put life within us, raising us up from death of self-gratification, and self-regard, and making us meet denizens of that eternal world in which man now dwells, though sightless, senseless, and unparticipating; thinking amid the universal bounty how much he himself can get?

CHAPTER II.

OF THE WORLD.

Covetousness which is idolatry.

WILLINGLY we admit that the universe is more than is embraced in our conception of it; that there is unknown being in that which surrounds us with these inexpressible enchantments. The infinitude of space made awful by unnumbered worlds, the things that subserve our use and pleasure, the forms of wonder and of beauty that life puts on all around us, truly these are more than we know, they are higher than any thought of ours. But surely they are. When I am looking at the sun is there not the sun that I am looking at; when I touch a tree is there not the tree? How can I look at or touch a thing if it does not exist? Men are not to be talked out of their most necessary and certain convictions. Reason does not avail against consciousness.

This is right and good, a most necessary position to be maintained. If men could have been induced to give up this conviction, they could never have known that which is of the utmost concern to them to know; never have learnt in how true and absolute a sense man wants life, or have discovered the fact that in truth exists, and why they are affected by it as they are.

It is certain that the things that are perceived

by sense exist to us; no question can be raised respecting them that does not touch ourselves also. If they are not, then are we most woefully deceived. For not only are we compelled of necessity, and without any reflection of which we can be conscious, to feel convinced that they truly exist, but all the circumstances connected with them agree with this conviction. We perceive them or do not perceive them, or we perceive them with variations of appearance, precisely in conformity with the results which should ensue from their existence. We can act on them and produce effects according to ascertainable laws; effects which react upon ourselves: whether there be fire or not, at least it burns us if we touch it. Or if we deny the existence of these things, what foundation have we for affirming anything? Of what then can we be sure? Not of our own existence, for that is inseparably bound up with our perception of these things; not of anything of which consciousness informs us, for its authority has received a fatal shock.

This is good argument. It has been proved valid by the result, for it has convinced mankind. We cannot put aside our natural conviction respecting the world, leaving our natural conviction respecting man, intact. Yet an alteration of our natural conviction respecting the world is necessary: it cannot be avoided. Nature cannot be that which we have necessarily supposed it to be. Examination proves those natural convictions false, and we know they cannot be true, because they have been formed without the requisite means of judging.

Especially is it demonstrated that our impression respecting the world, necessary though it be, cannot be a true one, because it has been found that it can only be maintained by forbidding inquiry, and by the assertion of a sovereign certainty in the impression itself. It should need no other evidence to prove the falsity of any opinion than that it requires such a basis. Yet we may see that it was necessary that this position should have been taken, and that men should have asserted that our idea of the world must be true, because with our amount of knowledge (that is in our ignorance) we were obliged to believe it. Contradictory as it is to all the principles by which man attains deliverance from error, he could have thought in no other way. For to think otherwise demands that he should recognise deadness in himself. The question which is raised respecting nature cannot be solved on intellectual grounds alone. It is also a moral question, it touches of necessity the spiritual. To ask: why that which is not the fact is the fact to us; why we perceive and feel the world as we do, if that which we so perceive and feel be not the very fact that exists; brings us into the presence of the profoundest religious questions. Life, death, eternity, all things divine and of deepest moment, have a stake in the decision. Till man know that this state is not his true life, he cannot allow that the world which is so real to him is not the true reality. Argument appeals to him in vain. He replies: 'I feel it, my consciousness declares it is so.' Nor does it avail to point out that every false opinion has the same grounds to sustain itself upon; that all appearances are necessarily believed true until increasing knowledge explains them: he will not believe himself deceived until he recognise that he wants his true life. Nor should he. The two convictions must come together. For man to believe that he feels that to be which is not, and at the same time not to see that there is a fatal defect in himself, ought to be impossible.

Yet that the argument for the true existence of the things perceived by sense is of no validity may be easily made evident. It rests on a confounding that which it is necessary for us to infer, with that which is true. It is doubtless necessary for us to infer many things respecting the world when, with our amount of knowledge, we reflect concerning it. Among these necessary inferences seems to be this, that the universe exists in the way that it appears to exist, or having such qualities as we feel it to have. But this inference gains no certainty by virtue of its necessity: it may be the result only of our ignorance. Nay, it is not even in strictness necessary; for all the impressions upon which we found the inference might be produced in other ways, and unknown conditions of our own would necessitate our perceiving in a way that should not correspond to the truth.* Evidently

* It is commonly argued, that from the feelings we have, we necessarily believe the existence of things which correspond to our conceptions. This may be true: indeed it is true until we accept the idea of a defect of being in man, which causes things to be to him other than they are. But the argument itself is strangely suicidal. For if our having certain feelings makes our belief of these things necessary, then it is clear that we should have that belief equally, in whatever way those feelings were produced. And we know that those feelings might be produced in other ways than by the true existence of the things we are thus made to believe; in dreams for example. Even if we may not say that the existence of those things could not produce the feelings, it is certain that there cannot be shown the least adaptation in them to do so. But, in any case, to show that our belief in the existence of things corresponding to our conceptions necessarily follows

the argument for the existence of things as they appear, or as we have necessarily believed them to be, rests on the conviction that we are not under illusion, and has no other ground whatever. But whether we are under illusion or not is the very question to be solved. To be under illusion is only to be ignorant of some essential circumstance. It is affirmed that we are under illusion, that this is right and natural, and proper to our state, and that it is of the utmost moment to us to understand that we are so, for thereby we learn our own condition. How should there be a reluctance to admit that we have been under illusion? The discovery of that fact is always the best thing that can befall us. Ignorance means being under illusion; all advance in knowledge consists in the discovery that we have been so. If this illusion, that the universe is such as it appears, be the greatest of all, differing from all others in being not intellectual merely, but affecting our life, our self, our very being, then is the discovery of it the greater gain. A knowledge that we have been under that illusion, and the escape from it, were a better thing than the escape from all others. It

---

from our having certain feelings takes away all the proof of their existence which those feelings might be supposed to give; if we account for our having those feelings in any way, we account equally for our necessary belief. Is it not better to say, that something not corresponding to our conceptions is the cause of our feelings, and that we necessarily believe as we do because our conceptions are necessarily inadequate? The inadequacy of our conceptions of nature is denied in asserting the authority of that necessary belief of ours. We ought from our feeling and our necessity of believing, in our ignorance, as we do, to infer the existence of something different from that which we conceive; something more: something above our power of conceiving, and therefore made less to us than it truly is, by virtue of our own state of being.

were to have a juster knowledge of the very being of the world in which we are.

What a light it throws upon our life, what a harmony it introduces into this tangled and discordant scene, that we should be feeling wrongly; that the universe should not BE that which it is to us! The mystery of the world is gone. If we feel wrongly, and think wrongly, putting that which seems for that which is, then do we understand ourselves. 'Surely man walketh in a vain show.' Is not that the secret of this strange life of ours, that might be so beautiful, so Godlike, but will not be: the raging passions, the vainly striving will, the expectation strained to the uttermost, to end in discontent, the hands for ever grasping, never full? The dark problem of humanity resolves itself in gladness, the universe springs up in light and joy. Man must have a different life, that which IS being the reality to him. Here will be the remedy for our ills; the cure for our diseases. Life poured into us: God's own Life flowing within us as it flows around: the glad current bounding through our languid veins, turning the coldness at our hearts to love, destroying the emptiness within, that strives to fill itself with vanity, by an eternal spring of blessing.

Man's life to phenomena, to things that are not, is his want of life That we feel this state to be one of life is no evidence that it is so. Who is so certain of his knowledge as he who most is ignorant? He above all feels and thinks he knows, and just as he gains knowledge does he become conscious of his want of it. Why should not he who most wants life feel and think himself most possessed of it? Why should not a growing life more and more make us conscious of want of life? When man's life is perfect, phenomena shall no more be to him realities as they are now, but as they truly are, phenomena only, forms and appearances of a different fact in relation to which alone his life shall be. We shall know that unknown essence, partake that now unapproachable EXISTENCE. We may understand what it must be to be in the eternal, in the spiritual world: that which IS shall be to us the reality of our existence, shall be to us then, as these inert phenomena are now, the facts by which our life and being are determined.

We naturally ask,—what are these things that we see and touch, which make the conditions of our physical life; our own bodies, and the external objects with which we are related by their means. No question is more important. But it is already answered. These things are phenomena; the things that appear. They are that which is to our sense and to our intellect. Brought into relation with the fact that truly exists, we perceive such things; being such as we are, and so circumstanced, we are impressed in this way. Admirable is the name which science has given them of phenomena. They are the forms under which the fact is perceived by us; the appearances, or things that are seen and felt, by the study of which we have to learn both what that fact is and what we are. We cannot affirm that they truly ARE, because that would be to deny that true existence which is above our knowledge or conception, and the existence of which is the sole reason that these things are perceived by us. Even so, to affirm that there is a *disc* in the heavens would be to deny the moon. This physical temporal world is the appearance to us of the world that IS, the eternal and spiritual world; and we believe it to be not an appearance only, but itself a true existence, simply because we do not know that true and absolute fact which causes it to appear. We are in the eternal world, and thus we feel it. We perceive the appearance to us of the eternal world, and call it the world that IS. But man is wiser than his own thought. He cannot rest in this belief. Science examining this world, which he feels and believes to BE, pronounces it but an appearance, calls it phenomenal, affirms that the true Being of it is unknown. Hence comes the difficulty, the logical perplexity, the necessity that men have been under at once of affirming this (which we conceive) to BE, and yet that the true existence of nature is not to be conceived. This is the apparent world, as at once opposed to and dependent upon the world that IS. As we understand that the appearances we perceive by sight indicate something different, which we conceive but cannot see; so we should understand that the phenomena we conceive by thought indicate a different existence, which we may KNOW, but cannot conceive. And as that which is but phenomenon, though by our defectiveness it is felt by us as existing, does not truly exist, so of necessity it cannot truly act. It differs from that which IS by being inactive. It is necessarily found to be inert. Thus a definite and intelligible difference is recognised between the phenomenon and the fact, and we are able clearly and consciously to distinguish between them. While the true being of nature is regarded simply as unknown, the phenomenon must be practically regarded as the fact. But to know that the phenomenon alone is inert, and that the fact is spiritual, entirely alters our conception of the world. The phenomenon takes it right place in our thought.

These physical and transient things are the mode under which we perceive the eternal. They present it to us. From them, first recognising the want of life in man, we learn what the eternal is. For well may these things that are seen teach us the unseen. Worthily do they fulfil their task. Image and symbol of love. Love, holiness, sacrifice, law perfectly fulfilled in perfect liberty, self utterly cast out: these are the fact of which nature speaks to us, which she images visibly before our eyes. These things are the forms under which the fact is perceived by man. The forms are in time, the fact is eternal; the forms are inert, the fact is spiritual; the forms appear, are felt, are conceived, the fact appears not, nor is to be conceived, but IS.

CHAPTER III.

OF IDELAISM: AND THE PROPER MEANING OF THE WORD MATTER.

We first raise a dust and then complain that we cannot see.
BISHOP BERKELEY.

IN nature, when one thing ceases another takes its place. For example: if wood be burnt, it is resolved into smoke and ashes. And these different things we regard as forms of the same essential existence. Thus it is easy to see the necessity of the conception of matter: while all particular things change or cease, there must be something which does not cease; something of which all these things that change are forms, a 'substratum,' which is the same in all. This conception, that the world consists of an unchanging matter, is a very obvious and natural one. It could not but have occurred to men, and have been commended to them by its apparent self-evidence and necessity. Nor does it seem easy to understand, at first, how the existence of matter should have been called in question, and have become the watchword of an apparently interminable strife. For the dispute concerning matter shows no sign of coming to an end. In spite of all attempts to close it, or to represent it as compromised, it is incessantly renewed. Men of science, as well as metaphysicians, descend into the arena.*

* See especially PROFESSOR FARADAY: *A Speculation concerning Matter:* and OERSTED: *The Soul in Nature.*

But this curious episode in man's history becomes quite intelligible, when it is viewed from the true vantage ground. We may see why matter must be asserted, why it must be denied; why the denial of it seems ridiculous, yet cannot be refuted; why the whole dispute appears absurd, and yet why men cannot disentangle themselves from it, or can only avoid it by refusing to think at all on some questions of the greatest natural interest and attractiveness. Nothing, indeed, can better illustrate the position of man in respect to the world in which he is than this very controversy. For the point on which it truly turns is whether the appearance of the universe corresponds to the fact; whether our natural impressions respecting existence do, or do not, require to be rectified.

If the fact be such as the appearance is, then there must be matter. Matter, therefore, is necessarily asserted, because the correspondence of the appearance and the fact is necessarily assumed, until by larger knowledge we are able to distinguish between them. It is a hypothesis to which we are compelled to have recourse, while we consider that which appears to be that which is. But, on the other hand, matter must be denied. It is in this way that the human intellect expresses its feeling that the appearance and the fact are not the same, that the universe is not truly such as it is felt by us. An expression imperfectly, and even inconsistently, made, because of defective knowledge, but not, therefore, without its value. The materialness of the world is asserted on the one hand, and denied upon the other; asserted, because the existence of a world such as we perceive it, involves the existence of matter; denied, because the existence of matter involves contradictions and untenable conceptions.

The question at issue is not one of existence, but of mode of existence; not whether the universe is, but whether it is such as it is felt by man.

A simple illustration will make clear the nature of the disputed point. When we look at a straight chimney through defective glass, the chimney appears crooked. And if we had no experience by which to correct our impressions, we should necessarily suppose it to be crooked; we should necessarily infer a crookedness. But in this condition of our knowledge, it might be argued on indisputable grounds that there could not be such crookedness; its possibility might be disproved. How, then, should we be situated; on the one hand, the evidence of sense affirming the existence of a crooked chimney; on the other hand, argument proving the impossibility of it? Just as the metaphysicians have been situated: sense, on the one hand, affirming the existence of a material world, argument proving that it cannot be. We should have found it as hard to understand that the dispute about the chimney affected, not its existence, but its crookedness, as we have found it hard to understand that the dispute about the world affects, not its existence, but its materialness. Nor could anything have solved the problem, but the discovery of what it was that caused us to perceive the chimney crooked when it was not. So can the dispute respecting matter end only with the recognition of the cause that makes us feel the world material when it is not. Meanwhile, it is interesting to observe that our mode of perception of the chimney would necessitate our inferring a property, or abstraction, of 'crookedness,' which had no existence, nor anything corresponding to it; for so we may more easily understand that conditions affecting our perception of the universe may necessitate our inferring a quality of 'materialness,' or abstraction of 'matter,' that has no existence, nor anything corresponding to it. Matter, therefore, is a hypothesis, necessary to be believed in so long as we think the phenomenon is the fact. It has been believed in, because the phenomenon is the fact to us; it ceases to be necessary when we understand why it is that our impression is not true.

Another illustration, perhaps more worthy of the dignity of the subject, may be found in the history of astronomy. So long as, through ignorance, that which is perceived by sense in the heavens was necessarily believed to correspond with the truth, it was also found necessary to suppose solid revolving wheels, by which the various heavenly bodies were conceived to be carried in their revolutions. These wheels were the epicycles; they were necessarily believed while the appearance was believed to be the fact, and ceased to be necessary when the truth was recognised, and through increasing knowledge the appearance was understood to be an appearance only. 'Matter' is like the epicycles in this, that it must be supposed while the appearance is believed to be the fact; ceases to be supposed when, through increasing knowledge, it is seen that the fact is different from the appearance. If the universe which truly exists be not inert, then the necessity for supposing matter, which arises from our perception of it as inert, is done away. Men have believed in matter because they did not know that the apparent inertness was due to man's condition.

The course of thought respecting it has been this. Matter is necessarily to be inferred if the world be such as it appears. But then if matter be disproved, if that inference be proved erroneous, it follows that the world is not such as it appears. This would seem to be an easy solution of the question. But here a difficulty interposes, for it is in any case hardly possible to give up the existence of that which appears, until we know what it is that causes it to appear. And to the majority of men, in this case, it is quite impossible. We feel the world to be such as demands the supposition of matter, and until we recognise man's want of life we do not know why we must feel it so if it be not truly so. This is why matter cannot be given up, why the denial of it appears like a denial of the world altogether, a contradiction of common sense. For it involves an alteration of our conception of man; a recognition of illusion, and of a cause for his being under illusion, which is contrary to our natural impressions.

The difficulty may be well understood by conceiving the epicycles denied without a recognition of the motion of the earth. It might have been argued, truly enough, that the epicycles could not exist. They are impossible. But the denial of them would have seemed to contradict perception, for it would involve the denial of the apparent motion. They can be dispensed with only on the recognition of that other motion which, involving ourselves, affects our perception. In a similar way, to deny matter involves the denial of inertness in nature, for if nature be inert, matter must be inferred. But this seems to be denying perception, for we do certainly perceive an inertness, nor can matter be given up, except by the recognition of that inertness which, pertaining to man, affects our perception. Briefly, the denial of matter is the denial of inaction in nature; it is necessary, therefore, because there cannot be an inaction in nature, matter is easily disproved; it avails not to disprove it, because the inaction is perceived. The only possible solution is to recognise the defect in man which makes him perceive it.

The parallel that has been indicated may be carried farther; for the mental life of man is emphatically one. As the complex and cumbrous hypothesis of the epicycles, felt at length to be impossible, was the means by which the earth's motion was made known; so are the complex conceptions, to which the hypothesis of matter compels us to have recourse, the means by which we are made to know the inertness, or deadness of man. Once recognised, indeed, these hard-won results (of astronomy in transferring the motion to the earth, of science in transferring the inertness, or defect, to man) appear self-evident and impossible to be doubted; they furnish so evidently the solution which the facts of man's experience demand; but the toil which wins them is long, the opposite conviction deeply rooted, and at first the difficulty in adopting the new conception almost insuperable; the admission that men can have been so long mistaken hardly possible.

The arguments against the possible existence of matter need not here be recapitulated. It is sufficient that they are allowed by almost all who have paid attention to them to be logically conclusive; so that the ground which is taken on the other side is a falling back upon consciousness and common sense, the affirmation that reason cannot deal with these questions, and that matter must be believed although it can be disproved. But the grounds on which the things that we perceive have been asserted to be ideas, are too instructive to be passed over. When the objects of sense are attentively investigated, it soon becomes apparent that there are qualities conceived as belonging to them, which are in truth inseparable from the mind by which they are perceived. Very obviously their colour and temperature are such qualities. Without a mind, there can be neither heat, nor cold, nor colour. These are sensations, as much as pain or pleasure. The same mode of reasoning proves all the other perceived qualities of sensible things to involve a mental appreciation. Hence the inference: that which is perceived by sense involves mental elements; it cannot exist except in a mind; but that which exists in a mind is an idea; therefore the objects of sense are ideas. So, by a series of deductions very hard to be escaped, we find ourselves driven to a conclusion the most incredible. For if one thing be more certain than another, it seems to be that the things which we perceive by our senses differ altogether from ideas. There is an externality, a substance, in the one, which there is not in the other. Whence then comes the difficulty? On the one hand, it is unquestionable that that which I call a thing cannot exist, as it is perceived, without a mind; yet on the other hand, I mean by it, and feel that I am right, a thing that exists independently not only of my own mind, but of all minds whatever. By the 'thing' I do not mean a state of mind, but that which causes the state of mind. There is here, however, nothing that is not perfectly simple. The idealist has but exposed an error on our part, showing that our premises lead to a false conclusion. He has argued from the natural assumption that the appearance corresponds to the truth of things, and by his conclusion proves the assumption wrong. By studying that which appears, we find that it is not, as we had supposed, that which exists; therefore that which exists is of a different kind. The basis of idealism is assuming the existence of the phenomenon, or that the fact corresponds to the appearance. For that which appears is proved to have qualities which show that it cannot exist independently of a mind. But this proves nothing about that which IS. Just in so far as the appearance is different from that which truly exists, so far must it depend upon the mind, and cannot possibly exist without it. For this is only to say, in a circuitous manner, that it is an appearance, and not the fact. The admission that the things that are perceived are ideas, or exist in a mind, can be extorted only so long as we choose to grant that the apparent mode of existence of the world is the true mode of its existence. The idealist therefore proves that the world is not such as it appears. But in truth his argument should not have been needed. That the appearance of the universe to us must differ from the fact is evident in the nature of things, and is confirmed by the analogy of all perception of individual things. The idealist, therefore, has proved to us that which we might have known before. He has reminded us how wrongly we are thinking.

It is interesting to observe the practical issue of this argument. Man's life is illustrated by it; for the question is not one of mere speculation, our profoundest feelings and beliefs are implicated in it. The idealist, of course, repudiates matter. For, proving that the things perceived, if having the qualities with which they are perceived, are states of a mind, there remains neither necessity nor possibility for that hypothesis. It is no more required. The substratum is mental, not material. It is for this reason that the controversy turns on matter. That is not indeed the true point in question, but it furnishes a convenient issue on which the discussion may be raised. Matter must be inferred if the idealist be wrong, need not be inferred if he be right. This form of the argument, however, has the disadvantage of giving to the question an aspect of abstruseness and unreasonableness which by no means rightly belongs to it; making it appear to be a question respecting existence instead of one respecting mode of existence.* For by matter, men in general mean 'things,' the world. They do not recognise that which Bacon terms 'the phantasmal matter of the schools' at all. If the question were put before men as it truly is, and they were asked: 'Can the universe truly be precisely such as it seems to us to be; must there not be much less in its mode of being to us, through our want of capacity to know, than there truly is? Ought we, or ought we not, confidently to affirm that something which no man knows anything about must certainly be, because otherwise the universe cannot be such as it appears to us; and ought

* In truth the question, instead of being merely speculative, is eminently practical. It is remarkable to note how the idealist argument has been, from first to last, subordinate to ethics. The idealist writers are primarily moralists, almost without exception, speculation being wholly secondary with them. And this is quite natural. Necessarily, the question of what the world truly is, is of all the most practical, and deeply touching the life and action of men. And that view of it which brings it most into union with our mental and moral being is especially adapted to a view of man's life from a practical and moral standing point. By no means is a man who says, 'The world can exist only in a mind,' a mere speculator. He says that, because he is resolved that his mind shall subdue and mould the world, and turn it to noble uses.

we to make this assertion in spite of an overwhelming argument against it?' there can be little doubt what the reply would be. Men scoff at the denial of matter, only because it implies to them the denial of any universe at all. They mean, by the word, something wholly different from that which is denied. Men in general are far enough from being idealists, but they are at least as far from believing in matter. They affirm that there exists a real world which is not an idea; but they do not mean to affirm that it exists in such a way as to involve opinions which can be shown to be not true.

For it cannot be too clearly understood, that by asserting matter we merely assert that the universe is such as it appears to us; we assert that it truly is inert. We will not be content to say, only, that we perceive it so, which alone we know. That seems not modest enough, not sufficiently distrustful of our own powers. For the due exercise of humility we prefer to say: The world is certainly such as it appears to us; but reason cannot solve the problems which arise out of our thinking so; we must believe things which are contrary to unanswerable argument. But in which of these courses lies the true humbleness, in which the wisdom, in which the safety? Which is best sanctioned by the course of Providence, which shows the truest trust and submissiveness? Which shall we think: that God suffers us to perceive the universe as it is not, and bids us learn that we do so by the exercise of the powers that He has given us; or, that He suffers us to be under the necessity of reasoning wrongly, and of contradicting by the best and clearest exercise of our powers that which is true? And this, with no result, no escape, nothing but a mere unsatisfied perplexity, or worse, a self-satisfied contentment? Looking truthfully into life, which is most like His other dealings?

For humility and genuine abasement of self, let any man understand that the universe is truly spiritual, and is inert to man only by his own want of life; there is a source of humbleness in that conviction, which need not be reinforced by any limitation of the possible achievements of the mind. When the intellect is in its right place, it needs no curbing for fear of pride: it has no longer any power to make proud. And, in truth, what men are prouder, more self-satisfied, than some of those who most insist on man's incapacity to know? That leaves quite untouched those relative superiorities on which pride is nourished. It is a vain conceit that high gifts conduce to pride: that man may be too much lifted up by feeling that his Maker has been too greatly bountiful to him.*

Taking our impressions and necessary conceptions as the standard of existence, it is clear there must be matter. If matter be not, man must be under illusion, and that not in respect to his opinion or thinking merely, but in respect to his feeling and being. There are therefore two alternatives. Either there is matter, or man is defective. Of these alternatives the latter surely is the more reasonable, and the more humble. For to adopt the former and say, there must be matter, puts us at once in the position of affirming a thing to be true on the sole ground that we cannot possibly be feeling wrongly, cannot be thus defective. But this

---

* See BACON's argument on this point in the first part of the *Advancement of Learning*.

would be to put it entirely out of our power ever to discover whether we are feeling wrongly or not. For if that be our case (and it is clearly possible), we can be made conscious of it only in this very way, of finding that our natural conceptions lead to impossible results. If we will not admit it possible that we may be feeling wrongly, that is simply to make our opinion on that subject of no value, to repudiate the possibility of escape from error, to throw aside God's great gift—the power to grow wiser.

The question concerning matter is not, therefore, as we are prone to think, a mere speculation, fit to exercise ingenuity, but having no practical value. The necessity that has tied men to it, in spite of themselves, has a wonderful significance. It cannot be laid to rest, because it insists on a solution. We are apt to say, there *is* matter; it is nonsense to deny it. Content, ourselves, with words, we would reduce all others to the same standard; having no sympathy for those who thirst for knowledge. But God, who guides man for His purposes, better knows what it is fit for him to do. He stirs his heart with an unquenchable resolve not to leave contradictions unprobed, not to succumb beneath difficulties, or give up where he cannot see. We want more love, more feeling for those who feel differently from ourselves, less disposition to despise that for which we have no taste. It is a poor and miserable notion, that the opinion we are content to regard as final ought to content all men, that the mysteries we accept should be accepted by others too. Would we check and limit life, and place a girdle, measured by our own capacity, around the widening future of our race? To him whom God is training for eternity would we say:

think for ever in this way; it is the best conclusion I can come to? Do we call this being humble, and knowing the limit of our powers?

To say, merely, there is and must be matter, is not to ignore the controversy, but only to take one side in it, and that confessedly the weaker, so far as argument is concerned. To ignore it would be to have no opinion whether there is matter or not. And, indeed, that would be truly a wise thing on the part of those who refuse to study the question. What faculties have we by which we can so certainly know that there is matter, especially when the more we use our faculties the more doubtful it becomes? The question truly is, whether a particular inference, a certain hypothesis or mode of accounting for our experience, is necessary. We, taking for granted our natural impression, say it is; the idealist, examining the facts of the case, says it is not. Why should we be so anxious to maintain a hypothesis, and, above all, a hypothesis which explains nothing? For we permit ourselves to rest in an idea that the supposed existence of matter explains our sensations. But if we reflect, in what way could the existence of matter explain sensation, into what remotest shadow of connexion can the two things be brought? The chasm between matter and sensation is impassable. The mystery of our consciousness is only made greater by the supposition of matter. We have asserted the authority of an impression, a natural belief; so far, doubtless, we have gratified a tendency of the mind, and have a certain satisfaction, but we have explained nothing. We have only fixed in irremediable confusion and darkness every question that can arise respecting the nature, the reason, or the mode of our experience;* a confusion which extends itself to all other questions whatever, except those which have reference merely to the relations of phenomena. For we have affirmed the existence of that which appears, instead of making use of that which appears, to learn from it that which exists. Never can we answer any question respecting existence, till we have rectified that error.

Matter is affirmed, simply because we do not see any other way in which our sensuous perception, and our mental feeling, could be such as they are. But how weak an argument is this, even at the best. Must a particular supposition be true, because we cannot otherwise understand how certain events should be? Must we be able to account for all things? Do we ever reason so, except as conducting to a mere presumption, diminishing in value precisely as our knowledge of the whole circumstances and possibilities of the case diminishes, and never to be maintained for a moment against the least sound argument to the contrary? Let it be granted that there is no other way in which we can account for our perception and consciousness, must we, therefore, assert a disprovable proposition? Should we call this common sense in any other case? Should we not say simply, and at once: we do not know enough? Let it be granted that the world is not an idea, and that it is proved also that it cannot be matter; is there nothing else that it can be? Do ideas and matter exhaust all possibilities? Is there no

* See the many and even extravagant theories that have been made to account for it on the supposition of matter. The 'Pre-established Harmony,' especially.

other way in which our experience may be accounted for?

There is another way. The very difficulty itself opens another way to us. There is the belief, the proof of which is hereby given to us, that man is defective, and that he feels as facts, or as having true existence, things which are merely appearances, and do not exist:—that his feeling is deceptive. Not having the life to know that which IS, he has felt and believed that to be which is not. Not knowing his own condition, he has assumed a defect in nature instead of in himself. Thus it is we are compelled to infer matter, and, when we have done so, find we have done wrong. Arguing from the premiss that the phenomenon exists, we necessarily arrive at a false conclusion. An evident solution of the mystery is here. There need not be matter, though we have been obliged to infer it, if man feel wrongly and have been under illusion; and that he has been so, is the key to his whole life.

Thus, understanding our condition, we perceive why idealism must arise, why it must fail. It must arise, because, through our false feeling, the inference we draw must be false, and idealism shows it to be so; it must fail, because idealism only proves the error, does not remove it. Idealism also rests, like the belief in matter, on the assumption that the appearance is the fact. One error vitiates both the opposing schemes. Instead of the true world, which is spiritual and eternal, our natural impression gives us a world temporal and inert, and idealism gives us a world which exists only in thought. Neither will do. Each refutes the other; strong to destroy its rival, impotent to maintain itself. The conflict was indispensable for our deliverance;

invaluable good has arisen from it, but also this small evil, that men think because idealism fails, therefore the opposite opinion must be true; that if the world be not an idea, then it must be matter. Not perceiving that it is precisely on the materialness that the idealist bases his argument, and that to assert it is to concede to him his own ground. Matter is the phenomenon, it is that which we conceive; if that which truly exists have the qualities called material, then is the world an idea. The world is not material apart from perception.*

If it be granted to the idealist that that which we conceive is that which exists, he is at once victorious, for that can only be in a mind. This, indeed, is evident; that which is thus conceived is the phenomenon; we want to know what that is which causes the phenomenon to be perceived by us. The idealist asserts that it must be the very thing that is conceived, or something exactly corresponding to it, therefore also something that exists in a mind, therefore an idea. The true answer to the idealist is, that the fact which exists has not those properties which he points out in the phenomenon, as pertaining to the perceiving mind. The fact is different from the phenomenon, and

* There is no mystery in this: it is only a peculiar and indirect mode of expression arising from the particular nature of the argument. From another point of view, we might speak quite intelligibly of ordinary things in the same language: *e.g.* apart from perception the moon is not bright, nor the sun revolving, nor even are leaves green. Colour as being a sensation, of course is only in being perceived. It is a power of causing that sensation that we mean to affirm of the leaf. Of course it is only another form of the same proposition to say: apart from perception there is not a material world;—apart from perception there is not a revolving sun.

must be, because in the latter are included negative qualities, which preclude existence. In a word, the fact is spiritual, is eternal, not to be known either by sense or intellect, which are in relation with phenomena alone. But for this answer the way was not prepared; it involves a recognition of man's want of life. Therefore it was argued that the fact did, indeed, differ from the phenomenon, but in some unknown, undefinable way; that the matter, which is asserted to exist, had not any of the properties which are perceived, but is a substratum wholly unknown. This would be good, so far, if it were genuine. That the fact is unknown is the best thing that can be said, until something be known about it; but the way in which this unknown substratum, still called matter, is maintained, is in effect a subterfuge. By being affirmed unknown, it is withdrawn from discussion; while by its being called matter, the concession is virtually done away, and it is practically endowed with the properties of that which is perceived. The maintainers of matter yield to the idealist reasoning in words, but not in thought; they adapt their words to resist the demands of argument, but leave the false conception unrectified. This also, however, could not have been otherwise. In truth, the doctrine of an unknown substratum denies the existence of that which is perceived, and asserts a mere unknown existence in its place. On this ground the idealist attacks it. He says: that which exists is this known, felt world, with these qualities which it appears to have, and by virtue of which it can exist only in a mind; by asserting an unknown matter, you deny the world in which common sense makes us believe, which is the world that we perceive and know. Here is the exact difficulty of the question. We wish to maintain the true existence of the phenomenon, but find we cannot do so. In one of two ways we must deny it, either by allowing that it can exist only in a mind, which is to be a phenomenon, and not a true existence; or by asserting that that which exists is absolutely unknown, which involves that that which is perceived, and therefore known, does not exist; again meaning that it is a phenomenon, and not a fact. There is truly here no difficulty or paradox whatever. All the appearance of it arises from our necessity of maintaining that to be truly real which is real to us. When once we recognise that that which is real to us is not truly real, and therein become aware of our own state, there is no more any difficulty. The existence of the phenomenon ought to be disproved; the endeavour to maintain it ought to make manifest that it cannot be. Our reason does not fail in dealing with the question, does not land us in paradoxes; rather it proves itself precisely suited to the inquiry; it elicits the very results that ought to be elicited, proving by the untenable character of the conclusions the falseness of the premiss. If inquiry failed to prove that the material things we naturally suppose to exist cannot exist, then our faculties would be deceptive and inadequate; they would be unable to expose our error. It is remarkable that the very truthfulness of our faculties, and their power to grapple with the problem of the world, has been the basis of the opinion we have formed of their inability: because they led us to a result we were unwilling to accept, we said that they deceived us. We would not be guided by reason, and give up the reality of that which we feel as real; we would not believe what it implied respecting man, therefore we said, reason cannot guide us. Not that men should or could have done otherwise. The world goes the course which God has appointed; the process of thought cannot be cut short. But we can do otherwise now. We can understand why all these various opinions and disputes have been necessary. We can trace them all from that defect on man's part, whereby he feels the phenomenon to be the fact.

And especially to be admired is the assertion, against the idealist, of an unknown substratum. That is, above all things, necessary. It is the assertion that there is a fact, a real world, though we may not know it, against the assertion that there is nothing but the phenomenon. Rightly, in one sense, has this argument assumed the name of common sense. It is the unquenchable feeling in man that the world, in which he is, is a real, actually existing world. 'If these things that I perceive be not existing, save by perception, the world exists notwithstanding, though I may not be able to know it. That which only exists in being perceived is not that which I mean when I speak of the world, and which I am sure exists, without which, indeed, nothing could be perceived at all.' It is the actual, eternal, not-inert world of which this is spoken. This is the world which must, and does, exist; is the cause of all perception, of all experience, yet which cannot be known (not known that is, by sense or intellect); with which all our experience brings us into continual relation, with which alone we truly have to do, but which is not the things which change and pass, is not such as that which we perceive by sense, or conceive in thought. This is truly the doctrine of common sense. 'There is an unknown EXISTENCE, which I can neither see nor think, which constitutes the true being of the world, and is the cause of all my consciousness, of all my perception. How different soever that which appears to me may be from this which IS, whatever may be proved about the former, however impossible it may be shown that it should exist as I think; all this cannot affect the being of the latter. Necessarily that which appears and that which IS must differ; if I have thought otherwise, it could only be from want of reflection. In this respect, at least, I know they must differ; that which appears to me is inert and transient, that which exists cannot be so.'

But it has been a great error to give to this unknown existence the name of matter; alike a great abuse of words, and a palpable confusion of thought. In one sense, it has the effect of rendering argument about matter impossible. As an absolutely unknown substratum, it is of course absurd to deny matter. The word stands merely as a symbol for that which exists, be it what it may. But for that purpose the word matter is not suitable, for it has a previous meaning, from which it is almost impossible to detach it. And that there is a delusion in this use of it is evident. By matter, the very men who call it absolutely unknown mean a thing with certain properties. Otherwise to call the world material were unmeaning. By this use of the word it is sought, at once, to maintain the existence of that which appears, and to escape from the consequences of that position; an attempt to ward off the force of an argument, as if it could be a loss to us to have our errors made manifest, a gain to hide from ourselves the falsity of our conceptions. Meanwhile, two results ensue. First, that the word matter is not available for use. It has no fixed meaning. If it be merely that which IS, and be entirely unknown, then it may be synonymous with spirit on the one hand, or with idea on the other, and there need be no question whether the world is material. To say that it is material is merely to say that it is unknowable, which had better be said straightforwardly, and in a manner less liable to be misunderstood. For it is, in fact, a great and invaluable truth that the true being of the world is in one sense unknowable: it is not such as can be thought, for that which is thought is inert. The spiritual is in this sense unknowable; that which can be thought cannot BE.* If, on the other hand, as is evidently the case, by terming the world material, it be meant that it has certain qualities, if by matter be meant that which occupies space, is hard and heavy, then the force of the idealist argument is not evaded, and all that has been said about the substratum being unknown is labour lost.

And secondly, this abuse of thought is severely avenged. For from this contrivance of calling matter unknowable, arises in great part the idea of the limitation of our faculties and inability of our thoughts to grapple with the problem of the world. Because we have chosen to assert the existence of the phenomenon, and are therefore driven into contradictions when we attempt to reason, (as we ought to be in order to deliver us from that false assumption), we seek to shut up the universe from human thought, neither entering in ourselves, nor suffering those who seek to enter in. That is our remedy for the results of our own errors:—not

* It can only be to us, or relatively. It is phenomenal. This is involved in the admission that our conception of the universe is inadequate; different therefore from that which IS.

to think! We say, nature is essentially mysterious; as if God had mocked us with a world the reality of which we could not reach, and, gratifying all other desires, refused to gratify the desire to know. The intellect need take no pains to limit itself; it meets no insoluble problems, for it can know phenomena, and phenomena alone are presented to it. Existence does not come within its sphere: Being is known in another way. To know that which IS, is to know God.*

But the word matter, in the abstract sense, seems capable of receiving a distinct and suitable meaning; matter might be defined to be 'a substratum necessarily inferred on the supposition of the absolute existence of the phenomenon, independently of a mind.' That is, if that which appears to us is held to be that which exists, and that man's perception is not modified by his own condition, then 'matter' must be supposed. If in

* Yet is it not a truer humility, a more genuine affirmation of the limitation and weakness of our faculties, to say that man must not assume the authority of his own impressions; but must inquire and learn, and remember his own defectiveness? The 'voluntary humility,' which imposes limits on our inquiries, vainly endeavours to compensate for the pride that lies at its root. Bold and aspiring in its commencement, lame and impotent in its conclusion, is the scheme which asserts the existence of that which man feels to be. It overleaps itself and falls on the other side. The assumption that we do know brings its own punishment in the necessary consequence that we cannot know. Suppose we invert the order; put the humility first. Is not that nature's order? Let us say rather, We do not so authoritatively know what IS; we have not faculties to know in this intuitive manner; we must examine, and use means, be willing to give up and learn. Then we find no limits; never need despair, never waste our ingenuity on that vainest of all 'anticipations,' of how much man can know.

this meaning it be asked whether matter exists, evidently it does not; for that which exists is not such as appears, but different; and that which appears *is* modified by man's mode of perception. The hypothesis of matter is inapplicable. We have been compelled to believe in it, and to maintain it against all the arguments by which it has been assailed, because, by our own state of being, a false conception respecting the being of the world has been made necessary to us, and the arguments against matter did not point out what that state of our own was, by virtue of which we were compelled to infer it.

We do, and must, however, use the word matter familiarly in a different sense, applying it, not to the unknown existence, but to the phenomenon; to that which appears, or is to sense and to thought. This probably is the best application of the term; for if it be used otherwise, there unavoidably arises a most embarrassing discordance between its common and its philosophical meaning, and an appearance of mystery in the question which does not rightly belong to it. Men in general always mean by matter, not an unknown substratum, but that which appears, the things they know. If by the assertion of matter different persons mean opposite things, it is no wonder the question seems insoluble and absurd. The entire confusion comes out of retaining the name of matter for the unknown existence, a sudden and violent alteration in the meaning of words. Matter should rather mean always the phenomenon, which is perceived by sense and conceived in thought, not that which IS.

The true difficulty in dealing with the question of matter is rather moral than intellectual. Idealism seems to check the sympathies, to cut off the basis of the affections, and leave no real men and women in the world but each man's self. This feeling has in fact a true foundation. The work of idealism is to show the unreality of that which we feel to be, not to reveal that which is. We feel rightly that it leaves a blank, an emptiness. But this is not by virtue of what it denies, but of what it asserts. To deny the materialness of the world is tolerable enough; it is the assertion of the idealness of it that cannot be allowed. To deny materialness is not to make less, but to make more. Idealism fails, not through giving up matter, but truly through keeping hold of it. The idealist asserts that the things to which we give the name of matter, the passing things we see, and touch, and use, are the true facts of the universe; that there is no fact in it but these things. Denying matter in words idealism asserts it in fact; while the opposing doctrine, which asserts under the name of matter an unknown existence, in truth denies it. Idealism affirms of the universe that it is such as the appearance is; the assertion of an unknown substratum affirms it different.

We deny that God is material, and feel that He is thereby not less to us, but more. His existence is intensified, rendered more absolute, more real by that denial. How should that be if to be material were not a derogation from Being? It is even so: to be material is to be wanting in being;* therefore we are obliged to deny materialness of God.

* It is in fact to be phenomenal, or apparent; to have no existence except in a mind, or in being perceived, to adopt the circuitous language of idealism; materialness involves inertness, defect.

By doing so, we affirm of Him a truer existence. How then should it seem to have a contrary effect to deny materialness of the universe? There is here a manifest inconsistency in our thoughts. In truth we thereby deny that the universe is wanting in being, we raise it to a truer existence. God does not sink into an idea by being proved not material, nor does the world that he has made. We should bethink ourselves here: it is not hard to understand how, by affirming nature to be material, we degrade it and make its existence less than it really is. The present difficulty about matter does not stand alone. Man had the same embarrassment respecting the materialness of God, as he has now respecting the materialness of the world. And it was removed also in the same way; namely, by his recognition of his own defectiveness and the necessary inadequacy of his appreciation.

As to the forms which idealism has taken, a few words will suffice. Berkeley, who for clearness and profoundness of thought has perhaps never been surpassed, argued that since the things that are perceived can exist only in a mind, and yet are evidently independent of our own minds, therefore they exist in the mind of God; are, so to speak, ideas in the Divine Mind, which He causes us to perceive according to certain laws dependent on His will. Evidently this is not to deny the existence of the world. It merely gives it a mental 'substratum,' instead of a material one, which should be held rather to make it more than less, to elevate rather than to degrade.

This conception has been variously modified, especially by the Germans, who have to a great extent represented the world as dependent upon human thought. The most remarkable form of this subjective idealism is perhaps that of Fichte, who asserts of nature that it is but the limit of the personality of man. But however extravagant any idealistic schemes may appear, it should not be forgotten that the root and foundation of them all, a root from which they must inevitably grow, is the affirmation of the existence of the phenomenon, or that the universe truly is such as it appears. This assumption contains and is responsible for them all; they must spring out of it, because, from that premiss, sound reasoning and argument inevitably conduct to them. And by rigorous logic and reasoning a certain number of men will always be guided, be the conclusions what they may. The idealist systems are necessary to deliver us from our false conception by showing what it leads to, and they can only be terminated by the fulfilment of their task. They may yield to an actual, spiritual universe, as ghosts disappear before the light of day; but they will never succumb to the assertion of a real 'matter.' When the true fact is recognised, and the reason why it appears to us as it does, then idealism may be heard of no more, but it must last till then; for repulsive as it is to some, it is not devoid of charms. It has its beautiful and admirable side.

There is again what may be called a scientific idealism, by which matter is rejected, and nature is considered as force alone. Under this view, particular objects are regarded as forces in space, variously modified and mingled, and no other substratum than the space is recognised. Unquestionably this is, for scientific purposes, an admirable view. Forces in space are the things with which science has to do. It knows nothing of that supposed matter of which the properties alone are to be perceived. But it is evident that force and matter are inseparable. The separation is excellent as an expedient, and adopted as it is by so many of the leading minds engaged in physical research, it exhibits strikingly the tendency of science towards non-materialism; but it cannot be a final solution of the question.*

The sum of the entire idealist controversy may be thus expressed: There is not a material world, such as we are conscious of perceiving, but there is a truly existing and real world, different from that which we are conscious of; and the reason we perceive it as we do is man's defectiveness, through which he is conscious of defect as if around him. The problem has taken long to solve, but it is not peculiar. Many men have co-operated to do that in respect to the world, which each man does for himself, many times in the course of every day, in respect to individual things; correcting his natural impressions by discovering that things cannot truly be as they appear to him.

---

* The ordinary idea of matter is that which occupies space. Substance is solidity. Yet there does not seem to be any essential connexion between the idea of material existence and space. The Divine Omnipresence is not felt to involve materiality. Still there is an obscurity in this subject, for the further discussion of which see Book V. Dial. I.

CHAPTER IV.

OF SCEPTICISM: AND THE GROUNDS OF KNOWLEDGE.

Our errors oft do aid us.

THE evident failure of the idealist theory to satisfy the demands of consciousness seemed to throw men back upon a belief in matter; but with this difference, that whereas, before, that belief had been supposed to be warranted by the reasoning powers, then it was recognised to be, at least to a certain extent, in opposition to them. The true existence of material things was to be believed on the basis of consciousness, and in spite of certain logical deductions from premisses that were still to be maintained. From this position, two things necessarily follow: first, that consciousness must be affirmed to give certain knowledge; and secondly, that our faculties must be held incompetent to discuss the true nature of the world. For, it is argued, if consciousness do not give us certainty, how can we attain certainty at all? and if our faculties be not incompetent, how is it that they contradict that which nevertheless we must believe?

This is the present position of philosophy. It is engaged in a struggle to maintain the authority of consciousness. But can this task be achieved? To assert the authority of consciousness is one thing; truly to maintain it, another. It is difficult to see what argument can be had recourse to, at last, but this, that God would not let us be under illusion.

Against which it is replied, that our conviction on that point is no proof; and the weakness of the philosophical position is rendered manifest by the fact, that men do continually deny the authority of consciousness in the very respects in which it is asserted to be undeniable.

We need to be on our guard here against the tendency, that is in us all, to rest satisfied in an opinion that suffices for ourselves, without considering whether it suffices for others also. When we are willing to admit a certain conclusion as necessary, and as the end of a strife to which we can see no other termination, we are apt to conclude that others ought to do the same, and to hold it a small matter that all are not convinced. But opinions satisfactory to certain individuals, or to certain classes only, ought not to content us. We require opinions that shall be evidently true, not to some, but to all; reconciliations that shall command all suffrages. If philosophy cannot attain to these, she fails as yet of her mission. She has to enlarge her grasp of things. A true philosophy shall do no violence to any instinct or aspiration of our nature; shall not leave men divided into hostile classes, adherents of opposing faculties.

Beautiful it is to see how nature provides for the advance of knowledge, by the different mental characters of men. Some attribute the chief value and authority to one portion of their intellectual being, others to another portion; and so it is achieved that no portion shall be finally disregarded, or defrauded of its share in the determination of human thought. For every system which, as all imperfect systems must do, asserts a scheme of things at variance with the full exercise of any of the mental instincts or capacities, is sure to be balanced and held in check by an opposing system. Uniformity of thought can be the fruit of truth alone.

Philosophy must reconcile the conflict between the consciousness and the reason, not by sacrificing one to the other, but by uniting them. Nor is this truly a difficult task. For the demand of consciousness is not to be believed, but to be explained, or accounted for. No theory can be accepted by mankind that puts consciousness aside; we demand that whatever explanation is given of anything, it should show satisfactorily why our consciousness in respect to it should be such as it is. To assert more authority in consciousness than this betrays a want of consideration; that would be to claim for ourselves an infallibility extremely hard to establish, and certainly alike needless and unsanctioned by experience. For what error is there that might not have claimed the warrant of consciousness, as the term is now used, until it was known to be an error. Is not every necessary belief based on the evidence of consciousness? But the beliefs that are necessary to us vary with our knowledge. Nor is security attained by claiming authority for universal belief alone. Not to speak of the extreme difficulty of ascertaining in what opinions all men do truly agree, such universality only proves a common condition; causes of error affecting all men produce universal errors. The motion of the earth is an obvious instance.

The authority of consciousness is a refuge from scepticism, a dogma laid down for fear there should be found no authority at all; but happily the danger is illusory, for the refuge is vain. If consciousness had truly possessed authority, it would never have been necessary to assert it. Least of all so to reiterate the assertion as has been found needful in recent times; a truly efficient authority does not need constant proclamation. The best interests of man, the interests of morality and religion, receive a grievous wrong in being staked on the authority of consciousness. Consciousness can bear no such burden: it is too weak, too fallible, as indeed the result has proved. And what are they to do, whom the best exercise they can make of their reasoning powers conducts to conclusions opposed to those which we regard as the deliverances of consciousness, and who think that such exercise of their powers is a better guide? Are their toils and aspirations to be nothing to us; shall we have no sympathy to give, no counsel to receive, are we only to say to them, you must think as I do? Why has God made them to think differently from us, but that we should mutually impart and receive?

Consciousness claims to be accounted for, not to be believed. To account for our consciousness is the object of all inquiry. All acquisition of knowledge, all discovery of truth, consists in learning rightly to account for our consciousness, or for our perceiving as we do. It is the universal problem; the work of science and of philosophy alike, which encounter the same task in different departments, to show what the true relations of things are, and why they must impress us as they do.

That consciousness demands to be accounted for, and not to be received as authoritative, may be made evident by familiar instances. We perceive, or are conscious of (for the terms are used interchangeably in respect to this argument) light, in the sense of luminousness, as existing around us. Yet we do not believe this; we account for our consciousness by the existence of motion or of something equally distinct from luminousness, external to our eye, and are content; we feel that all the demand made by our consciousness is satisfied. It is the same in innumerable other cases. In truth, the idealist argument is based upon the authority of consciousness, the reply consists in an attempt to account for it. For the idealist maintains that there is truly luminousness and not motion around us; he asserts that that which causes his consciousness must be such as his consciousness testifies. But the assertion of an unknowable matter, against idealism, accounts for our consciousness by the existence of something very unlike that which consciousness testifies. And that this is the better method is proved by the fact that the unbiassed sense of mankind gives it the preference.* In truth, we can only be satisfied when our consciousness is accounted for; the mere assertion of its authority always leaves a dissatisfaction. It divides men into two camps; some asserting its finality, others reasoning against it. And it should have this effect, for to assert authority in consciousness implies imperfect knowledge, which implies wrong conception. When our knowledge is perfect on any subject, we can always account for our consciousness. To claim authority for it, while unaccounted for, is to convert ignorance into an argument. Consciousness alone is necessarily defective; it can vouch only for our feelings; it cannot embrace the truth of things, but only the mode in which we feel them. From the same facts

* The defect of the idealist argument is, that it does not account for our consciousness; it builds upon it as a basis, and does not show it necessary. To account for anything is to show it necessary. To show our various states of consciousness to be necessary is the entire work of inquiry.

different beings have different consciousness; a blind man must be conscious of darkness as the condition of the world, but his consciousness is accounted for by his defect of sight. What consciousness vouches for, in every case, is that we have a certain experience and nothing more.

There is no ground for the fear that, if our consciousness be not authoritative, our demand for certain and trustworthy knowledge cannot be fulfilled. We imagine this only because we have been in too great haste, and could not suspend our opinion and inquire. The feeling indeed is natural, and is confirmed by the result of past endeavours to attain certainty upon other principles. Yet it may be observed that the appeal to consciousness also fails to give certainty; its evidence is diversely interpreted, disputed, denied. And the past attempts to arrive at certainty have failed for the very reason for which the appeal to consciousness also fails. They are indeed but appeals to consciousness under another form. They agree in taking ourselves, our impressions, as the standard of being, and in affirming that which is to us to be that which IS. None of them seek, by exploring man's own state, to ascertain in what respects that which is to him differs from that which truly is.*

But how evident is the omission we have made; how simple to repair it. The light which is thrown on the past vain struggles to attain certainty is perfect. That which is to us cannot be that which is, because our own state of being must have a

* This may be demurred to on behalf of Kant and others, who regard time and space as modes of thought. But the statement seems to be justified: the question relates to man's being, not to his modes of thinking.

share in determining what shall be to us. This we have overlooked, and have struggled, therefore, necessarily in vain, to frame to our thought a consistent or intelligible scheme of things. It is no paradox, no difficult or unusual mode of thought, that the things which are to us, or are real to us, are not truly real. It is self-evident and necessary. And this is the whole secret of scepticism. That form of opinion must have existed, ought to exist. Scepticism is but the denial that that which is to us truly is. There is nothing evil in this, it is true. It is but the bringing out into view that which is involved in our own principles, but which we try to conceal from ourselves. Scepticism is not an enemy but a friend, and does us apparent mischief only through our own falseness and mistake. For do not we ourselves say that the essence and very fact of being, or of that which IS, is not to be known or thought by man? Why then do we find fault with him who puts our principles into a practical form, and says: this which we know and think is not that which is? There is no more in scepticism, in its relation to the intellect, than this: our own most necessary and valuable principles applied. If scepticism be morally harmful, we must remember that it has its root in our inconsistency of thought, and that it lies in our power to do away with these effects by ceasing to be inconsistent. We look on scepticism as if it were some monstrous tendency in man to deny that which is true, to refuse certainty when it is legitimately offered. But it is not so; there is no such tendency in man: no man loves doubt for its own sake. Men deny that that is which is to us, repudiating an authority in our feelings; they deny that we can truly know by sense or intellect.

But herein they speak truly and not falsely; we ourselves say the same thing. That which is to us is not.

We need not, therefore, fear scepticism: it is evil only because we have placed knowledge on a false basis. We do injustice to ourselves, and think too meanly of the Creator's bounty. For it is only our false persuasion of the true existence of that which is to us, that embarrasses us with paradoxes, and makes us think our faculties too poor for our desires. Truly our faculties must be mistrusted, our aspirations checked, if our reason, in denying that which consciousness seems to teach us, were denying that which ought to be believed. The doctrine of man's incapacity for dealing with any problem that can be presented to him has no other foundation than this. It is based, therefore, not on a true humility, but on pride and self-confidence; on our firm presumption that that which we feel to be must be.

The problem that is truly submitted to us to solve is to account for our consciousness; or, from that which is to us, to ascertain what truly is. Surely this is a simple and in no way intractable problem, and one, moreover, in respect to which all the previous exercise of our faculties, all the course of human observation and thinking, prepare for our success, and supply principles for our guidance. We need only to apply the established and evident rules of inquiry according to the nature of the case, remembering that it is a question of being, and not of thinking, of life and not of intellect. By study and observation, by patient watchfulness, and suspension of opinion until the materials for judging are acquired, remembering that we do not naturally know that which is, but have to learn it humbly, we must first ascertain what elements in that which we feel to be are due to our own condition; what qualities or modes of existence, in that which is perceived, cannot truly belong to that which is. These will show us what our own condition is, and the fact then will stand distinctly before our apprehension in its true nature. Our consciousness will be accounted for. We shall understand why the fact, being such as it is, must affect us as it does. And scepticism, at least, will be no more a foe.

We are apt to think that if consciousness do not give certainty, if we have not true knowledge directly from its evidence, then we can never know. But the contrary of this is the fact; if we were thrown on the direct affirmations of consciousness, then we could not have true knowledge, but only a presumption more or less strong; which indeed is proved by the language held by those who maintain its authority. That consciousness requires to be interpreted, is the very reason that we can have certainty in our belief. Consciousness, though it does not give direct knowledge, affords the elements of knowledge, a basis for absolute conviction, to be worked out by the conjoint exercise of all our powers. Consciousness, and reasoning, and conscience or the moral sense, all have their necessary part in the determination of that final question, What is it that EXISTS? Man must give all his powers to that work, must consecrate to it his whole nature. He errs grievously when he strives to snatch a fallacious certainty by any shorter process. He must submit to curb his impatience. Scepticism avenges the partial exercise of his powers; it means that there are elements in his nature which have not received their due weight in the determination of the question, faculties which have not contributed their part. Scepticism is a symptom, not a disease; demanding not to be itself suppressed, but that the disease be cured.

Nor can it be suppressed; it must be embraced and turned into a support. Denial cannot be cured by denying, it must be swallowed up in affirmation. Evil must be overcome with good. The sceptic says in effect, 'That representation may suit its purpose very well, but what is to be said of this which I find in myself, and which is in opposition to it? I do not believe it. There is more in the world than that view takes in.' Nor can he be silenced, nor ought he, except by the enlarging of our hearts and intellects to embrace all that is found within humanity as the foundation of our thought, so that there shall remain nothing to oppose: no disharmony, no violence. Then scepticism must cease, it can be thought of no more, its spring is cut off for ever.

Most instructive, also, is it to note that there is virtually a twofold scepticism, answering to the two parts of man's nature which are set aside by our doctrine of the authority of consciousness. There is a scepticism of the reason and a scepticism of the conscience, theoretical and religious. Some speculative men will not cease to amuse themselves by denying the existence of external things on the ground of logic, and use such arguments to ward off appeals which may be unwelcome. Some religious men will insist that this world is a shadow, a dream, an illusion, not truly real at all, and that the only world that deserves our regard is unseen and future. Both feel that there is that within them which the argument from consciousness ignores and tramples upon.

The innocence and usefulness of the sceptical argument are evident, when a general view is taken of the problem with which our faculties have to deal. We have a certain state of consciousness, and a certain perception of things apart from ourselves. These are the elements from which we have to gather, so far as possible, a true knowledge. For this work we have three powers; sense, intellect, and conscience. The problem evidently is to harmonize all these constituents of our being, to fulfil them all: to learn why our sensations should be such as they are, in a way perfectly conformable to reason, and embodying also the demands of our moral nature. For this purpose we must pursue two chief inquiries: What appears to us? and why does it appear to us as it does? or, what is our own state, and in what way does that state modify the impressions we receive from that which is apart from us? Our own state must be learnt by the observation of that which we perceive; for, as we are, so must our perception be. It cannot therefore be the fact, that that which appears to us can be that which truly is, because our own state modifies our perception. Therefore, from the appearance, or the phenomenon, which we know cannot BE, we have to learn that which IS, by setting aside, if we can detect them, those qualities which are due to ourselves, or to our own state of being. Practically, this inquiry brings us to two results: first, we find ourselves feeling wrongly, for that which appears is felt by us, and thought by us, to be that which IS. We do not, except by an effort of reflection, disentangle ourselves from this illusion, and when we recognise it to be such, still the feeling remains the same. By this we know that there is a want or defect in man; for the phenomenon, which is not and cannot truly be, is real to us, affects as real that which we feel to be ourselves. We recognise a defectiveness in man, but this is no strange conclusion. It is not less self-evident than it is demonstrable by observation.

And when we look again, to ascertain what elements in that which we perceive, or that which is to us, should be attributed to our own condition, a truly marvellous simplicity appears. For we discover that certain of these elements are negative, denoting defect or absence. We perceive an absolute absence of true action. This negative element in that which is perceived, therefore, we know at once must be due to ourselves. For two reasons: first, that we have already recognised in ourselves defect, which must cause such appearance; and, secondly, that the very fact of its being negative renders it impossible that inaction should belong to that which IS. We take, therefore, this negative element in that which appears to be due to ourselves. And now the problem is in one sense solved. We have attained a general conception which is appropriate to the case, have placed ourselves at least in a right attitude. We know that the negative qualities, in that which we perceive, do not belong to that which IS; we know that there is defect of being in ourselves. Herein all our nature receives its full satisfaction: sense, and intellect, and conscience. Our consciousness is accounted for. We understand why our perception and feeling should be such as they are: why a world that has not negative qualities, a world of true BEING, spiritual and eternal, should make us perceive a world inert and wanting in being, a world the existence of which can be disproved. The intellect recognises a satisfying cause of our perception, and a satisfying reason why that perception should be such as it is: the conscience feels that it is true. The two portions of our knowledge, relating to ourselves and to that which is apart from us, are brought into a just correspondence. Why that which is not must be felt by us to be, we know; for that which we feel to be must have qualities which cannot pertain to that which truly is, qualities that are opposed to being. We understand that there must be less in that which is to us than in that which is; that the world must be perceived as having defect in it; that it must be to us physical, because it is felt as inactive. We understand how things that pass away, and cannot therefore be truly said to BE, are realities to us; how our existence is in time. We understand why unquenchable aspirations in our hearts make us long for a higher, worthier state of being; why our nature is at strife with itself, why we cannot but look forward to a future different from the present: why all thought has been perplexed; why we have been trying to conceive negations as existing, the phenomenon as fact; why we have failed: whence the belief has arisen that we must ever fail, and that our faculties are insufficient for our instincts. For we have thought that that which we feel to be corresponds with that which exists, not regarding that defect in man which makes us feel as we do; and our reason, therefore, denies the being of that which we feel to be, our conscience asserts a being above any that intellect or sense can reach. We see that all the past experience of man must have been such as it has been, leading him by the only possible way to discover his errors, to recognise his true relations, and to comprehend the meaning of his experience.

Thus scepticism is absorbed, and turned to good account; conducing to certainty in thought, and piety in feeling. It is not possibly dangerous any more, for there is nothing in us that is over-ridden or oppressed. There is no strife, because no violence is done; no tension, because no coercion is exercised. For the doctrine that is asserted is axiomatic and self-evident, the rebellious reason finds all its demands anticipated. It is granted that that which has properties which disprove its being IS not: that which IS is that which truly acts; it is that which must be. Its properties are those which cannot be denied of being, because they are involved in the very meaning of the term. No man denies that something exists: that being IS. That were a contradiction in the very words; if there were not something that exists, there could be no phenomenon, nothing could appear. Of this existence, then, it is asserted, that it has the necessary properties of being, that negative properties do not belong to it, that it is not transient, or in time, that it is not inert: that, therefore, all our experience, inasmuch as it must be caused by that which IS, and cannot be produced by that which is not but only appears to be, must be caused by that which is not in time but is eternal; and is not inert but is spiritual: that this is why that which IS can neither be perceived by sense, nor grasped by thought. Sense and thought deal with that which is inert and in time, with that which appears; but that with which we truly have to do is not that which appears, but that which IS, and to know which is not an intellectual but a spiritual state. In this, the demands of the reasoning faculties are fulfilled; these are self-evident axiomatic truths, on which we can rest with a solid and unwavering assurance. And the more, because the higher faculties of our nature find herein also a warrant and repose, which they cannot otherwise know. No task of argument and inference is laid upon them, ungenial to their nature. The longing for God meets its response. Not afar off, not to be realized by great stretch of thought, not separated by innumerable existences which intervene between Himself and us, but close around us, nothing between Him and our inmost souls, the BEING with whom we have to do. Not to be perceived by sensuous eye, nor conceived by thought, still sensuous, but known within the heart. And seen visibly, as alone His very Life and Being could be made manifest to us, in Him who first showed to man, unknowing, what it is to love.

If it be held that consciousness alone (and even in opposition to some other tendencies of our nature) can give certainty, surely much more should it be granted that the consciousness, the intellect, and the moral sense, united, should afford ground of certainty. We assume that they never can be united, because all attempts to unite them hitherto have failed; but for this failure there is a sufficient reason in the fact that hitherto regard has not been paid to our own condition of being in its influence on our perception. An essential element in the treatment of the problem has been left out.

Again: if consciousness authoritatively testifies to anything, surely it is to the existence of that which we directly perceive by our senses; to the existence of that which is to sense. But that which is to sense differs, in innumerable particulars, from that which is to thought or to conception, and we arbitrarily endeavour to make consciousness vouch for the latter. There is a twofold weakness in endeavouring to assert, on the authority of consciousness, that to which it does not directly testify, but which rests on inference. Why should we any more claim to base on consciousness that which is to our intellect, than that which is to our sense? We give up the latter, willingly, at the bidding of inquiry and sound reasoning, why not also the former? No scepticism results from giving up that which is to sense, but rather greater assurance every way. We give up colour, as existing apart from us, and think of motion, and feel nothing lost. Why should we not, equally without loss, give up motion, and believe something else?* The denial of that which is to sense is the life of science; might not the denial of that which is to intellect be the life of a true knowledge? It is, indeed, already denied in words, in the doctrine that the essence of things is unknown.

Or, again: if it be not denying the authority of consciousness to deny that which is to sense, and substitute for it that which is to intellect; if consciousness be not held to testify to the appearance to sense, but only to that which may reasonably be considered the cause of it; then neither can it be denying the authority of consciousness to deny that which is to the intellect. Consciousness must be held to testify, not to the 'appearance to the intellect,' but to that which reasonably, and by sound inquiry, may be shown to be the cause of it. Therefore, to assert the spirituality of nature, denying that which is to our thought, is truly, and in the best sense, to assert the authority of consciousness. For consciousness testifies emphatically to the

* If the colour be subjective only, why not the motion too? If the eye alter that which exists, why not the thought?

reality of the world, to its existence; to affirm the reality against the seeming, is alone to carry out its dictates. For sense and intellect do not exhaust our faculties; there are other powers, other modes of apprehending and feeling besides these. And these other faculties have their authority also; it is their presence, in truth, and the assertion of their claims, that unsettle the deductions of the intellect from sensuous experience, and will not let the notions which we frame in that way rest in undisturbed possession. The claims of these other faculties have to be made good, and their authority duly recognised, in our idea of existence, before we can have peace. Therefore, in denying the best conclusions we have been able to frame; in landing us in perpetual doubt; in overthrowing every structure we could set up; our faculties have not played us false, have not proved their incompetence. They have established their competence rather; hitherto, at least, they have proved themselves efficient. They have established a claim to confidence; for they have saved us from a fatal, though earnestly desired, security and repose in error. We wished to draw our conclusions before we had examined sufficiently, to affirm without proof, to take that which seems for that which is; but the faculties God has given us refuse. Too faithful to be lulled into indolence, or crushed into subjection, they have toiled in their appointed task: through good report and evil, they have done their work; and shall stand justified yet, to the confusion of our selfish pride, which has sought to pamper our own indolence, and justify our mistrust, by depreciating God's bounty to mankind.

It is of no avail to say: God cannot have suffered us to be under illusion, and the world must be such as we feel it to be. That argument is suicidal, it refutes itself. For the proof that we are under illusion is not avoided by that view. Inasmuch as our reason, when exercised upon the subject, leads to a result which we affirm to be false, we are still under illusion; deluded in respect to our intellect; our powers do still deceive us. The illusion is of a different kind, but it is not less illusion. And, in truth, what else can the doctrine of the incompetency of our faculties, and of our inability in our present state to solve the problems that are presented to us, mean but that we are under illusion, and think and feel erroneously; that the impressions we receive from that which is do not correspond to the fact, because of our defectiveness? The statements differ in form, and in the ideas associated with them, rather than in their true significance. It was right and necessary that men should have spoken as they did; with the problem unsolved, and evidently baffling all attempts to solve it, what could be said, but that their powers were unequal to the task? They had not then learnt enough, they had not the means essential to success: science had not revealed her secret. The history of the inquiry confirms the interpretation.

And what are the facts of our present life, which seem so marvellous and unaccountable, but the very experience which alone could correspond to, and express, the evident conditions of our being? Our feeling, our consciousness, our existence and action in this physical and temporal world, what are they but the only way in which the phenomenon, that which appears and is not, could be real to us? Our perception, our immediate and necessary conviction of the existence of that which we perceive; its substantial, solid, unquestionable reality to us; its correspondence with our powers, feelings, activities, desires; its influence over us, and response to us in every way; what is all this but the necessary feeling which must arise in respect to the forms which are in time, when the fact, or that which is eternal, is unknown and unfelt? In what other way could forms be facts to us? And that they must be so is involved in the universally admitted statement, that the true essence and being of the world is unknown. The mystery of this existence, of our consciousness, perception, necessary belief, no more exists as it existed before. Let man be conceived wanting in the true being by which alone he can be conscious of the true being of the universe, and his present feeling and experience must be the result. He must be conscious in this present way; feeling that to be real in which, when he examines it, he finds reality cannot be.

Above all, there must be in nature that unutterable and inexplicable mystery, that infinitude of wonder and significance, speaking to us of things so much above us, so much above anything we can find in the things that seem to us to be the facts; that strange disproportion and unmeetness between the life of nature and its substance, which disappoints us so when we strive to penetrate the secret of its glory, throwing us back in mere amazement, and with a doubt within our hearts which we cannot but resolve to quell. There must be that conflict in our souls, that strife between our most assured convictions and resolves, and the poor conclusions of the sense-bound intellect, taking the seeming for the fact. For what means that doubt, and dissatisfaction, and impossibility of reconciling our convictions, but that the conceptions we form of the world do not answer to that which we feel, are not adequate to the effects which it produces upon us. The power that is in nature overweighs and makes ridiculous our conception of a dead material substance. That which acts THUS on us surely must be active, not inert. Truly. But why then is it inert to us?

CHAPTER V.

OF POSITIVISM: AND THE RELATION OF SCIENCE TO PHILOSOPHY.

Assailed, but not enthralled.

STILL it may be asked, why should we not rest in the natural idea of a real existence, of which the properties are inertness and resistance, with a capacity for motion, such as we think of under the name of matter, and which we seem to understand so well until we are called upon to put the opinion to the test? Why should we not believe that such an existence has been created, even though we cannot conceive how it can be, and find ourselves driven into contradictions when we argue respecting it? Is it not better to remain in such an assurance as that acceptance of our natural ideas may give us, and to adjust our conceptions of spiritual things, and our belief respecting a higher existence, to that view of the world; pursuing rather questions of a practical bearing, in respect to which we can attain definite results? Why should our thoughts be unsettled: if these things are not truly as we feel them to be, is it not better, nevertheless, that we should believe them to be so?

A fatal objection lies against this compromise. It cannot be carried out. Men refuse to be bound by it. Nor can any means be found of giving it a practical effect. He who argues against the unflinching pursuit of truth cuts away the basis of all argument, and man's best instincts take part against him. Whatever the truth may be, it must be better to know it than to be in error; it must be a sacred and pre-eminent duty to accept it. Even if we can have no knowledge, it is better to know our necessary ignorance.

This is Positivism: the denial that man can have knowledge rightly so called. Positivism accepts the result of the argument which lands us in universal denial, and seeks to adapt man's actions to that opinion of his condition. It says: 'All human knowledge is relative; it is knowledge only of that which appears, not of that which IS. All things that we can perceive or think are phenomena, not truly facts or existences. The relations of that which appears we can know, but deeper we cannot penetrate. Nor need we wish to do so, for these relations of phenomena are all that concerns us, all that in any way affects our well-being or our duty. Man's life, in short, is a life that has to do only with appearances and their relations; it is not his part to inquire concerning existence.'

Positivism thus denies that the truth of things answers to our impressions; in this respect well representing the tendency of science (whose name it especially assumes) to exclude ourselves, and any mental necessities whatever of our own, as a standard of that which is. It lays down the principle, that that which is to man is not that which is, and repudiates any inferences founded on the true existence of that which he perceives. Of necessity positivism denies matter; for it denies that the world that exists corresponds to that which appears. In other words, this world, which is material, exists only relatively to us; that which truly exists may be different. The phenomenon is one thing, the fact (or absolute) is another, but with the latter we have no concern. And this ground is taken by the positivist, for the express reason that it is practically the best. He lays it down, that through pretending to any other knowledge than that of appearances, man's powers are perverted, and his efforts misapplied. Pointing to the past history of philosophy, he says, only mischief has come from the belief in the reality of that which we feel to be, of that which alone we can know. Men have wasted their labour in pursuits that are necessarily fruitless, and have turned aside from the works which alone can truly benefit them. The remedy for the evils of the world is to know, and be content in knowing, that we have to do with phenomena alone, which are real only to man.*

It is vain, therefore, to assert that we should believe in the true reality of that which we conceive (the material world), on account of the benefit of that belief, when so large and so powerful a body of men assert that such belief is above all things mischievous, a chief source indeed of all our practical evils. There can, at least, be no end of strife in that assertion. It is met by the counter assertion that we ought above all things to know that the things we perceive are but appearances. In vain we struggle to make for ourselves a place of rest, in which we may cling undisturbed to our natural impressions, and hold fast to that which seems good and right to ourselves. We are driven to face the problem of our existence; we may not shrink from it. If a belief in the true reality of that which is real to us cannot be maintained by argument; if it have no valid basis in the supposed

* Positivism being in truth an extreme idealism.

authority of consciousness; if it be so widely affirmed to be pernicious instead of beneficial; what are we to say? Shall we shut ourselves up in a mere resolution to believe, isolating ourselves from the world's work, or shall we fairly look the question in the face; fairly, because unfearingly? For why should we be afraid to let go our conceptions? Is not God's world infinite, and infinitely better than we can possibly conceive?

And nothing can be more satisfactory than the result of the inquiry on positivist grounds. Positivism seeks to set aside philosophy, or the inquiry into that which IS, and to substitute for it science, or the observation of that which appears. By the study of phenomena and their laws, to predict the future, for the regulation of human actions, is the sole object which it permits to man. But the conclusion which is thus set forth has not been established. It is true that science consists in the study of phenomena and laws alone, and has no reference to the fact of existence; but it is not proved that its relation to philosophy is that of successor and destroyer. There is another part which it may take, and for which it may be better adapted; that of servant and renovator.

In bringing science and philosophy into relation, and marking the links which unite them, and determine their mutual destiny, positivism does an essential service; but it is not always given to a man rightly to interpret a relation he is the first to perceive. Science labours so strenuously in her work of observing phenomena and tracing laws, and achieves such triumphs in that field, that it is naturally long before it is perceived that she has any other or higher task. Science has been conceived simply as the instrument by which our understanding of phenomenal relations is to be enlarged, and our practical command over phenomena extended, by ever-increasing knowledge of their laws. But we have seen that there is another possible result of science, in addition to greater knowledge of phenomenal relations: that it may also teach us something of ourselves; and may show us that some condition perceived as apart from man should have its cause looked for within him. By the nature of science, as the study of that which is perceived, it has an essential adaptation to this result. And if this be so, if through scientific study of phenomena, we are made to know our own condition, and to understand that a quality, or mode of existence, in that which is perceived, does not truly belong to that which IS, then it is evident, also, that philosophy stands in quite a new relation to the problem with which it seeks to deal. The experience which demonstrates its incompetency in the past has no force in respect to the future, for the conditions of the problem are altered. It was before: from that which appears to discover that which IS, our own condition, on which the mode of our perception depends, being unknown. Now it is: our own condition being known, from that which appears to discover that which IS. The latter problem being as evidently possible, and within reach of the human faculties, as the former is beyond them. For philosophy has found her problem insoluble hitherto, not because man has not the requisite faculties, but because the conditions, or data, essential to a determination of it, were not at her command. Science supplies the indispensable element, for lack of which philosophy has trod a weary round in vain. It transfers the negative element, the perceived defect, to man;

giving demonstration of an inaction, which cannot belong to that which truly exists, in that which man is conscious of.

In truth, the relation of science to philosophy is very beautiful, and exhibits, in an eminent light, the life and mutual dependence which mark the progress of human thought. Man's strivings after knowledge, in all his tortuous windings and blind errors, are not mere idle waste, but form a mutually connected and balanced whole, no part of which is unnecessary, and which tends with perfect aim to the development of his nature.

For man's true work is that to which his instincts prompt him; to learn that which IS; to pass beyond and through the mere seeming, to the sacred fact of being. But in this effort he fails, and is baffled. Over and over again he fails, for he takes the true being to be what seems to him. He seeks the absolute; but this absolute he conceives to be, first, that which he can see, or otherwise perceive by sense, and then what he can think. He seeks an absolute that is phenomenal. Philosophy is embarrassed by the effort to conceive true being that has negative elements in it; real existence that is physical or inert, an absolute that can be conceived. It embarrasses itself with needless contradictions. How, then, can it be liberated? Only in one way; only by science. Man has to learn to distinguish the elements introduced by his own state of being into that which he perceives; for he naturally attempts, at first, to frame his belief on the supposition that that which he feels to be corresponds, in all respects, with that which exists. Hence his failure; hence the power of science to remedy that failure. For when the impotence of philosophy stands confessed, the direction of men's energies is altered. They no more seek, so earnestly or so exclusively, to know that which IS; they give themselves to the investigation of that which appears, to the study of that which is to them, to the tracing of relations, to the establishment of laws. They say: We can never know the very fact of things, we were mistaken ever to try. Meanwhile, with these very words upon their lips, they remove, under a guidance unrecognised, the error which made that attempt a failure. They make manifest that the phenomenon does not correspond to the fact; they give demonstration in what respects it necessarily differs, revealing so what the element in our own condition must be, which is affecting our perception, and to which we must have regard in all our thoughts respecting that which is.

Positivism, therefore, is partly true, partly mistaken; true in its basis, mistaken in its practical conclusion; for it fails to recognise an essential element in the case, overlooking the power that is in science to make us know more of man. It is inconsistent also, inasmuch as it takes no account of the significance of its own fundamental position. For while it asserts that all which we feel to be, all with which we seem to have to do, is phenomenon only, and not truly that which is; while it lays stress on the fact that this, which truly is not, is real to us, and affects us as if it were the only reality; it omits to note the remarkable fact which it thereby proves respecting ourselves. It makes no account of the fact, that that which only appears, and is not truly real, is reality to us; is not felt by us as it is. It passes by the fact that we feel wrongly, which yet it puts forth as its especial discovery. It forgets, that if that which is not real is the reality of our being, then we are under illusion in respect to our feeling and our life, and that we may and do know ourselves to be so. This great and striking certainty, amid all the uncertainty which it points out, it overlooks. We thank it for its revelation of a strange, and strangely neglected, truth respecting ourselves, which we will pursue to its true bearings, and turn to its right account. For positivism affirms that man's existence is only relative. But surely this is to deny that he has absolute existence: and what is this but to say that he has not true, actual life; to affirm him wanting, dead?

Positivism does not deny that there is true existence; that were impossible to one who allows that anything appears, or that there are phenomena. And it admits further, or asserts indeed, that this true existence is not identical with the world of phenomena of which, by sense and intellect, we know the relations and the laws. Most unjustly, therefore, were positivism charged with atheism; and it is in the farthest degree removed from materialism. But it evidently errs in stating that our concern is with phenomena alone. That which EXISTS must be that which truly acts; must be the only CAUSE. That which only appears can have no action. Think or feel as we may, our true concern must be with that which EXISTS. If we feel otherwise, then we are deceived, but the case is not altered; for a mere appearance, as it has no true existence, can have no true action. There is an evident misapprehension in the statement that we have to do only with phenomena; for if that with which we concern ourselves do truly act on us, then it is not only a phenomenon; if it do not, but only seem to do so, then it is not that with which we truly have to do. Accordingly there is an inconsistency in the language of positivism, marking this inconsistency of thought; for these phenomena, of which it is affirmed that they are not that which truly exists, are, at the same time, spoken of as facts, or as realities. The truth is, simply, that which positivism expresses but ignores; that we are under illusion, and feel that which is not, as if it were. Hence all the mystery, all the confusion. It seems to us that we have to do only with things which may be shown to be mere appearances, and not true realities; but the fact is not, and cannot be so. Under the appearance of these phenomena, our real concern lies with the truly existing fact. We are wrong in thinking the phenomena to be that fact.*

If we may apply to positivism its own language, we might say that the *phenomenon* is, that we have to do only with phenomena. That is what appears; such it is to our feeling and apprehension; but it cannot be the fact. And again: when the positivist argues that man can know only phenomena, and cannot know the true and real fact, it is obvious, in reply, that if his argument be correct, it is

* There is a curious parallel between the practical teaching of positivism, and the theory of idealism. The idealist, examining that which we perceive, and finding it not to be a thing which can exist apart from a mind, affirms that that which exists, exists in a mind, that it is an idea. The positivist finding that the things which we feel to affect us are phenomena, not true realities, affirms that our concern is with phenomena, and not with the true reality. But both are inconsistent; both deny their own affirmation in making it; both are misled by the fact of our perceiving and feeling not according to the truth. The idealist should say: These things, that can only exist in a mind, are not that which truly is; they only seem to be. The positivist should say: Our concern seems to be with phenomena alone, but this is an illusion: the truly important thing is other than these.

impossible for him to know what he affirms. For he can, at most, know that man appears unable to know the fact; he can but know that this is the phenomenon, that so it is to his apprehension. He cannot know that the fact is so, else does he know more than phenomena with their relations and laws. For man's relation to the fact of being is evidently not one of the 'relations of phenomena,' to which the positivist affirms our knowledge to be confined. The data necessary to prove that man can only know phenomena can never be forthcoming, for their existence would overthrow the proposition.

Positivism evidently makes too little of man. Recognising that defect of his being which cuts him off from true reality, it bids him sink to the level of that state; from which all his strife, and error, and vain, disappointing labour rather should confirm his hope to be set free. But there is much instruction in the system. Positivism proves, at least, that the denial of matter, the denial of the reality of the things that are felt as real by us, is not unpractical, does not lead to neglect, or the withholding from those things of all due regard. For the very doctrine which most emphatically takes this ground in theory, in practice devotes the most intense regard to the affairs of life. If this be the result of the merely sceptical denial of the reality of that with which the senses deal, how much deeper, more earnest, more worthy must be his heed to his daily life, who recognises in ordinary things, not mere material existences, nor bare phenomena with no deeper meaning, but the absolute fact of being, filled with all the worth of the eternal, than which there can be no other and no higher, and which are obscured and darkened to our apprehension only by the want of a respondent life in us.

For while the positive theory, in rejecting any essential existence in phenomena, gives great liberty to thought, and overthrows some inveterate and baneful errors, its benefits are purchased at too great a cost. Its gifts are treacherous; uttering words of honour it inflicts on science a deadly wound. For science lives by the pursuit of truth and of reality. So she grew to her vigorous maturity; so must she continue to grow, or she must languish and decay. No languid impulse to ascertain relations can feed with throbbing life her mighty limbs. The warm current of her blood congeals at that icy touch. The balancing of profit with no hope to be nearer God, is a sickness at her heart. Enthusiasm and belief, an assurance that there is a reality, a being, a life verily responsive to our appeals, in that with which we have to do, these are the secret of her strength; 'her liver, heart and lungs, whereby she lives.' Like Samson shorn of his locks, and delivered helpless to his foes, were science robbed of these. Mixed with errors, and false thoughts, and beliefs unfounded, and mistaking of phenomena for facts, have been these living powers; but they are living. Like all life, struggling towards objects unforeseen, in failure and illusion, the longed-for rest ever forbidden to her weary feet, each solid-seeming goal found to be but an unsatisfying semblance when it is attained, and constituting but a new starting point in the pursuit.

Because the true ends are not man's but God's. Because in all that seeming failure and delusion, God's ends are fulfilled. The true life dwells with him triumphant, rounding our restlessness with everlasting calm, swallowing up our sorrows in the eternal joy. There is no failure. The failure is phenomenon, not fact; that which we feel because we feel wrongly, and know not that which IS. While we go mourning, the heavens clap their hands, and earth rejoices. Nature palpitates through every nerve with infinite delight. To know is to be glad. The attainment of our ends, our success, our content, were life no more, but death; death undestroyed, the victory of that which is not Love, whose victory were absolute defeat.

CHAPTER VI.

OF MYSTICISM: AND THE USE OF THE INTELLECT.

The real is God's ideal.

IF that which exists be not such as we can conceive in our thoughts, how is it to be known? or at least how can the intellect have any part in the attainment of a knowledge in the possession of which it cannot participate?

The answer to this question is simple. The intellect is a means towards acquiring knowledge in the same way that the senses are. Not a ruler, but a servant. Our necessary conceptions, or thoughts, are not correspondent with the absolute truth of things, but they are elements from which that truth may be gathered. They are materials to be used. Even so the impressions we derive from sense do not correspond to the relative truth of things, or to the right conception of them; yet are they the means by which we ascertain that relative truth. The intellect contributes to a knowledge not intellectual, as the sense contributes to a knowledge not sensuous. We learn from our senses, by examining, and ascertaining the conditions which cause them to be affected as they are. We learn from our intellect, by examination likewise, and by ascertaining what the circumstances are that necessitate our having the conceptions we are obliged to form.

When the distinction is borne in mind between that which it is necessary for us to infer and that which is true, the part which the intellectual operations bear in human history is evident. Our thoughts and conceptions are to be interpreted by a knowledge of ourselves and of our relations. They are among the elements from which we learn what the truth must be. It is not truly an embarrassment that our necessary conceptions should be incorrect, and should vary; these are the very circumstances which give certainty and completeness to our knowledge. It is because the impressions on our senses do not correspond with the objects of sense, and vary with all our changing relations to those objects, that sense avails for us as a means of guidance and information. Even so, it is the inaccuracy and variableness of our intellectual conceptions that give them their value, enabling us truly to know. If any one adhered to the immediate impressions on his senses, as indicating the true relations of things, he would necessarily be altogether deceived; but that would be because he misused his senses. He would ascribe to them a false authority. Even so are we deceived when we regard our intellectual conceptions as conveying to us the truth of things. We misuse the intellect. We ascribe to it a false authority. Nothing can be simpler, than that intellect is to be used as sense is used: one question is ever to be applied to both: Why have we this sensuous impression; why have we this intellectual impression? Sense and intellect, alike, have to do with the true reality not directly, but indirectly only, dealing themselves with appearances.

How then are the intellectual impressions to be corrected? for the impressions of sense are corrected by, and made subservient to, the intellect.

To what is the intellect subservient? To the moral sense: to that perceptive, appreciative power in man which is not intellectual. Sense and intellect do not exhaust the faculties of man. If the intellect be not truly a knowing faculty, but have to do only with phenomena, it does not follow that man cannot know. There is that in man to which the intellect is servitor, even as sense is servitor to it.

But further: the senses themselves afford a means of correcting their own impressions, even where those impressions cannot themselves be altered; we learn through sense to interpret sensuous impressions which do not correspond with truth. So the intellect affords a means of correcting its own apprehensions, even in cases in which those apprehensions cannot themselves be altered. We may not be able to conceive in any way but one, but we may understand that such conception does not and cannot correspond with the truth, and why it cannot, and in what way the truth must differ. From things within the sphere of our sensuous examination we learn, in part, what those things must be which are beyond its scope; so from the investigation of things that are within the grasp of our intellect, we can learn in part what that must be which is beyond it.

That we can only know appearances is a conclusion we naturally form, when first we find that our intellectual impressions do not correspond with the fact. But this is a conclusion too hasty: the facts warrant no such inference. Our intellectual apprehensions do not truly represent the fact that exists, therefore they must be interpreted, and used as the materials for obtaining true knowledge. True knowledge is obtained not by intellect alone, but by the exercise of all man's powers on data and means of inquiry, of which the intellect furnishes its share. This is the legitimate conclusion: to be tested not by inferences or arguments beforehand, but by trial and experience. It is true that by the intellect we know only appearances, and not the true reality. This is no special disqualification or disability of man's, but belongs to the nature of intellect as such. It is not a faculty that deals with the essence of being. To KNOW, in that true sense, is not an intellectual thing.* Our perplexity has arisen from misapprehension; from the expectation of finding knowledge where it could not be found. The discovery that by intellect we are in relation with phenomena alone is none the less valuable because it is an axiom, and involved in the nature of intellect itself. It has wrongly been made a ground of discouragement in relation to our capacity to know. There is more in man than intellect. Our natural supposition, that we have true knowledge in our intellectual impressions, only answers to the supposition, equally natural, that we have true knowledge in our sensuous impressions. Man escapes from both errors in the same way; turning that which has been a source of deception into the means of a larger wisdom. It is man's nature to be deceived by the intellect, even as it is his nature

* This is very evident when it is considered what intellectual knowledge is. Not to insist on the idealist argument that only ideas can be in a mind, is it not evident that the BEING of anything cannot be in a thought or conception? Is it not evident that the intellect can contain only that which is inert? That which is to the mental consciousness cannot be active; that is, it can only be a phenomenon or appearance. The simplicity of this makes it seem abstruse. Is it not self-evident that there cannot be in my mind the very fact and Being of anything else?

to be deceived by sense. It is his nature, also, to escape from being so deceived, and to make the false impression teach him the true reality.

When any particular opinion, or inference, is necessary to us, the fact is not, as we are apt forgetfully to assume, that what we are obliged to think is true. The fact is, only, that we are obliged to think it; that such an inference is necessary to us. We continually apply this principle in ordinary life: why should it not be extended? For example, men have necessarily inferred the existence of a motion, which is the cause of sound and light; but the fact in this case is, that it has been necessary to men to infer these motions, not that they exist. The right question to be asked respecting light and sound is, why it has been necessary, from our experience of them, to infer a motion. All philosophy has been put astray through failure to perceive the true fact in cases such as these; nor has science wholly escaped the perversion. And, again, from our sensuous experience as a whole, we have necessarily inferred a world having the properties we call material. The question that truly arises here is: Why has it been necessary to us to make that inference? But all man's inquiries have been perverted to discussing the 'material world;' how it exists, or can exist, how it came into being, what its essential nature is. The error and perversion of thought here is almost too simple to need pointing out. The wrong question has been asked. The only fact is our necessary belief or inference. Metaphysics has been driven to madness in seeking for the essence of that which is but phenomenon. To ask the right question, laying hold of the only fact in the case, the necessity of our inference or belief, sets the whole tangle straight.

Why it has been necessary for us to infer a material world, we may perfectly understand:* but why, or how, a material world can BE, we should necessarily ask in vain for ever. It cannot be, its nature is contrary to being; it is phenomenal only. Just so with respect to light and sound. It is not difficult to understand how we, from our perception of them, must have inferred motion as their cause; our idea of space and matter necessitates it. But how motion should cause us to perceive light or sound is truly, and must remain for ever, a mystery insoluble. That is the phenomenon, not the fact; a necessary inference to us, not the truth. Most unjustly have we depreciated the human intellect, because it could not solve questions which were misconceived, and recoiled from mysteries, absolute and never to be solved, because of our own creating.

The fact of nature is not that which is to the intellect, or which we can think, and have been compelled to think; but not, therefore, has the intellect no part to play in making that fact known. The intellect subserves true knowledge, not as itself true, or giving truth, but as a means; as part of the phenomena from which the fact is to be elicited. Our thinking and conceiving as we do is a part of that consciousness on which knowledge is founded. We ought to have conceived respecting the world as we have done; to have found those inferences necessary which have been necessary. From these very

* Namely, because that which we perceive is not the very fact but an appearance; which, therefore, we necessarily find not to act, or to be inert, inasmuch as it is only an appearance. We infer therefore an unacting existence, which is the definition of matter, because we have not distinguished between the appearance and the fact.

conceptions and inferences we may gain the knowledge that we need, knowing them to be erroneous. Without our false conception, and necessity of inferring that which is not true, how could we know the truth, of which these very errors of feeling and conceiving are an essential part? They are necessary results, or constituents, of the very truth which we require to know. They show us ourselves. Surely we can learn truth from our errors? What else is almost all our experience? The truth respecting man and the world involves those very errors; they are the means by which that truth is to be made known. For to know the truth, even intellectually, respecting any given existence, does not demand that such existence should be conceivable or comprehensible by thought. To know the truth is to think rightly respecting it, to understand its relations. This knowledge the intellect can attain respecting the fact of being, the right mode of regarding it, a knowledge of our relation to it, a recognition of that which pertains to ourselves in our perception, a consciousness that BEING is not to be known by thought, but by LIVING. The intellect fulfils itself in taking its right place.

That which is to our intellect, though not itself the fact, is the basis and means of our knowledge of the fact. As our sensuous experience, or perception by sense, is the ground of a knowledge not seated in the sense, nor corresponding with the sensuous apprehension, yet in which the demands of sense are most perfectly fulfilled; so our intellectual experience is the ground of a knowledge not seated in the intellect, nor corresponding with our intellectual apprehensions, but absorbing, using, and interpreting them, showing why they must be such as they are. That which we must conceive is not, therefore, assumed to be that which IS, but we understand why our conception must be such, the demands of intellect being herein most perfectly fulfilled. The great error and embarrassment of philosophy has been the supposition that that which is to thought is that which IS; assuming the existence of the phenomenon. As well might we attempt to construct a science on the basis of the existence of that which is to sense. We are as if a person should endeavour to understand the world, assuming that the forms of things to his eye are their true forms. We need to do for the intellect what science has done for sense, put it to its USE. We should no more think of the world we conceive as being the true world that IS, than we think of a chair as having but three legs because we may see but three. As we reflect why we can only see imperfectly, so we should reflect why we can only conceive imperfectly.

From that which alone can be conceived we learn to know that which is, by a knowledge of ourselves and our relations; as in every other case we learn the truth from the appearance. For, knowing our own state, and how it affects the impressions we receive, the necessity of our inference has a new and more fruitful meaning; it no more leads us to suppose that the fact is what we must infer, but guides us to the reason which makes us so infer; guides us, therefore, to the truth, to that which, operating upon us, should affect us in that way. It is not that our thoughts or perceptions are different, but that we interpret them differently; from our intellectual processes we infer, not the absolute truth of their results, but of what kind that fact must be, which, truly existing, is the true cause of all.

It is true that the phenomenon cannot directly

teach us the fact, but it can teach it indirectly. For the phenomenon can teach us ourselves; it has an emphatic and perfect adaptation to reveal to us what our own state is. Among the relations we discern in it our own relations have their place, chief and most needful of all. And when our own state is known, the phenomenon ceases to be unadapted to make us know the truth. It is perceived to be that which it truly is, the best and only means for giving us that knowledge. By no other means could that result be obtained; the appearance is that which should and must appear. It is the sole means whereby the fact could be shown to us as it is. We, being as we are, could not know, or be brought into relation with, the spiritual fact of being, in any other way than this, perceiving it as material and defective; nor learn what it is, except through feeling the inert phenomenon as our reality.

There is, in truth, no reasonable bound to the possible knowledge of man, when that knowledge is pursued in the right way, and its nature rightly understood. In the phenomenon, or appearance to him of that which is, and in his consciousness or knowledge of himself, are contained elements which are evidently sufficient for a knowledge beyond any limits we can conceive. At first he naturally mistakes, because he assumes the phenomenon to be the fact. But this is an error time is sure to correct: he finds that what he had supposed to be the fact cannot be so. Then he gives up the hope of knowing the fact at all; but this despair also is transient: the necessary laws of the mental life re-assert themselves. That which has been human experience in the past repeats itself in the present under another form; and man's native instinct of seeking that which IS, beneath that which is to him, crushed for a moment, resumes its rightful sway. The intellect willingly abandons its usurped authority, and puts on the higher dignity of serving.

Thus thought is set free, and called into new activity. Having authority only in respect to the mode in which we are affected, only in respect to phenomena, it needs no limits or restraints to be imposed upon it. A variable element, as it were, is introduced into our regard, on which the stress of all intellectual difficulties is thrown and lost. Our necessary conception depends upon our state, which indeed is to be learnt from it; the fact is wholly untouched by any such necessities of thinking. We no longer seek to compel our thoughts, to make them correspond with that which other portions of our nature demand; we are no longer under temptation to violate the laws of thinking. What we must think depends upon what we are: it is not absolutely true; was never meant to be so; cannot be. Our conception is not itself true knowledge, but is the impression on us from which true knowledge is to be derived. The affections are not coerced, thought is not distorted; they are harmonized by a new element. The fact and our conception ought to differ; their difference is the condition of our knowledge, not a hindrance to it; it teaches us to know ourselves. Just as much should it be so as the shape of an object to our eye ought to differ from its true shape.* Just as necessary to our true knowledge is the inaccuracy

* If the visible size of objects did not vary with our distance, into what perplexity our movements would be thrown.

of the intellectual apprehension, as of the sensuous one. Just as truly is it in fact correct, such as it ought to be. To our conception the eternal ought to be temporal, the spiritual ought to be material. So it is known to be eternal and spiritual; even as the stars ought to be seen by us as feeble lights, and are known thereby to be rightly conceived as a boundless galaxy of worlds.

The advantage of this position is, that everything is made a matter of rational evidence. The entire question respecting the world is brought under the ordinary rules of judging, submitted to examination, opened to any and every test. Nothing is taken for granted, no step in the process set aside as not to be subjected to rigorous proof. No appeal is made to feelings as giving demonstration, nor anything claimed to be true, merely because of our intense conviction. There is no Mysticism in it.

For mysticism is the assertion of a means of knowing that must not be tried by ordinary rules of evidence; the claiming authority for our own impressions. Against the tendency of men to deny all that is not perceivable by sense, the mystic affirms another and higher faculty, which has authority in its sphere, as sense has in a lower one. Against one impression not to be questioned, he puts another; doubtless without power to maintain it, without valid basis, but not without ample excuse. The assertion of the true existence of phenomena, or of the inherent and ultimate authority of sense and intellect, not only gives an example, but institutes even a necessity for the complementary assertion of the mystic. The essence of mysticism, and its source, alike, are found in the doctrine that we should believe, on the evidence of consciousness, in the existence of an external world such as we feel to be. Open that door, and no form of mysticism can be shut out; all mysticism is in its roots embraced in the doctrine of the power of consciousness to vouch for any existence but its own, and that of some cause, the nature of which must be discovered by investigation. No mysticism is so mystical, none so fertile of confusion, as the assertion of the authority of our consciousness to establish the existence of 'matter.' That is the parent from which the whole brood derives its life. No mystic, how extravagant soever, does anything more than assert the authority of his consciousness for something that he cannot otherwise prove; nor asserts on that authority a position more absurd, more clearly disprovable. All the weakness and mischief of mysticism, without its redeeming features or excuse, are in the doctrine that there must be matter because of our consciousness. The mystic does but meet the asserter of matter with his own weapons, on behalf of reason, and religion, and humanity. He does but refuse to be morally stifled, when the very arguments of his assailant are equally available for his defence. To cling to the existence of such an inert world as is supposed to answer to our consciousness, and to refuse to the mystic the existence of anything else that answers to his consciousness, is a puerile inconsistency. We demand of him that he shall submit his consciousness to the test of reason and experience, and consider how far it may be determined by his own condition, but refuse to submit ourselves to the same demand.* The assertion of

* The case is not altered by the assertion that matter rests upon the universal consent of all men, or that the indi-

matter as existing, as anything else than an inference necessary to us through our defect of knowledge, is the worst of all mysticism; mysticism without its poetry, its beauty, its elevation of sentiment, its charms of imagination. It is to be mystical on the wrong side; to incur all its faults and penalties, and forego its compensations; to defy argument and reason, and the legitimate restraints upon our tendency to judge of things by ourselves, not as the mystic does for the sake of freedom, but only to make ourselves more irremediably enslaved.

For men, not understanding that the true world is different from that which it appears, and feeling themselves, therefore, bound to maintain the existence of that which appears, are driven to the very argument of mysticism in order to do so. Finding that the existence of that which appears can be disproved, is indeed disproved by the soundest exercise of our faculties, they assert consciousness as an evidence for its existence, not to be questioned.

---

vidual judgment is corrected by mankind. For in the first place the case is not so; and in the second, if it were, it would not affect the argument. There is not universal consent for matter. It has been denied by reflecting men from the earliest times of which there are records, and widely prevalent philosophies are based upon the denial of it, at this day. But still more, there is no consent at all in favour of 'matter,' but from a few metaphysicians. The world have no such conception, can hardly be made to understand it, and when they do, it seems to them absurd to the last degree. What universal consent testifies, is that there is a real acting world, and that it is not such as philosophers represent. But even universal consent would avail nothing to the argument: the mystic affirms that all men have the faculty he claims if they would use it, and besides, no man has, at last, anything but his individual consciousness to rest upon. What avails it, for stability, to enlarge a pyramid, if it be inverted?

Hence this argument comes to be ever at the mystic's beck. 'We only know by consciousness, or an intuitive conviction, the existence of an external world; that can be disproved, and yet it is the basis of everything. Therefore we must admit the existence, upon the same evidence, of anything else; it is of no avail to disprove it. We believe things that can be disproved, and refer their existence to the mere mysterious ordination of God, without meeting the demands of the intellect by any explanation of how the case can be.' This argument is perpetually used, nor can it be set aside, except by a mode of reasoning which sets aside equally the existence of the world. The door is opened by the assertion of the existence of matter for any and every superstition. The argument by which all superstition supports itself has been conceded. For, in truth, the belief in matter, the belief that the world that we feel to be is the true world that IS, is in the strictest sense a superstition. It is THE SUPERSTITION rather; the idol, or show, which we worship, in which we believe. All other superstitions cling about this, and suck their life from it. Our ignorance, our actual spiritual death, whereby the eternal is not to us, and the phenomena or forms are the realities, this is the source of all the superstitions of mankind; even as to know THE ETERNAL is their remedy.

Mysticism is the result and necessary complement of the assumption that the world is such as it appears. The unconscious protest of humanity against the violence thus done to it. It is an effort to fill up the vacancy and defect of being in that which we feel to be. This is the good side of it. It recognises the defect in man's present state of being, and claims for him higher faculties than those that link him to the phenomenal. It refuses to make its belief of the spiritual a mere deduction of the reasoning faculties from that which appears to sense; proclaiming, more or less decisively, the unreality of that which passes. But it is vitiated by the present imperfection of our apprehensions. It does not embrace the true relation of that which appears to that which is. It sets up an antagonism in man's nature, not perceiving the needful union of all his powers in the work of learning the fact from the phenomenon.

In giving up our assumptions, and approaching the question of our life and the world with the free exercise of all our powers, recognising deadness in man, instead of defect in nature, the demands of the mystical impulse are fulfilled, and its errors done away. The disharmony of our thoughts and feelings, from which it springs, no more exists. There is no longer a chasm, which we must try, at least, to conceal, between our notions of the world, and that which our life as men makes necessary to our hearts. The fact to which the intellect testifies suffices for us, and needs no inventions to supplement its defect. For at the root of mysticism lies the feeling of the defectiveness and evilness of matter; a feeling that cannot be banished, strange as it is. We conceive God as creating a substance that puts limits to His power, checks and confines even our capacities, lowers and degrades the spirit. And man's heart and reason, alike, revolt at the supposition. The wrongness of the fundamental thought expresses itself in chimeras and vain imaginations innumerable. The radical perversion in the conception of the world gives birth, of necessity, to unceasing efforts among men to rid themselves of its intolerable consequences. Endless fancies and wild frantic theories, succeed one another vainly to that end; and must succeed, until that fundamental fancy that the world is such as it seems to us—that wildest of all theories, of a substratum conformable to that appearance—be exchanged for a recognition of that which makes a world in which is no defect, no unreality, appear to us, and be felt by us, as this world is felt.

For how can that world be real which might be an illusion, which we have in truth no means of distinguishing from an illusion? Surely a true reality could not be such as that; could not be a thing which an illusion could perfectly and undistinguishably simulate. What a mockery of reality were that. For be the fact what it may, it cannot be denied that our feelings might be exactly what they are, without the existence of any such world as we believe in. Let any man recall the phenomena of dreams; let him even think how he can know that he is not dreaming now. Are not certain and undeniable illusions as real to us, in sleep, as any part of the experience we call our life? the things we feel and see in dreams as real to us, as potent over our feelings, as veritable sources of joy or pain, as anything we dignify with the name of reality, and think it madness to deny? Truly we feel these things that are around us; they are real to us, there is no trifling with them: but even so are dreams; we cannot disregard their power, or treat them with unconcern.

> For in that sleep of death what dreams may come
> Must give us pause.

Dreams may thrill with acutest anguish, rouse to maddening terror, overthrow the reason; making trivial all things besides. A dream may be to us more, and more real, than all our waking life. How can that be a true reality which an illusion can outweigh? What do we need but to awake out of sleep, to be roused from our delirium, to know in very truth that we have mocked ourselves with visions, have only seemed to be moved by things that had no being; the true cause of all that we have felt being quite other than that which we have thought?

By signs and proofs unnumbered God warns us of our error; calls us to reflect and see, and be no more deceived. By our instinctive feelings, which cry out for a truer, more real world than this, by sickness and dissatisfaction of heart which prove its vanity, by the result of reasoning which demonstrates our misconception, by doubts, and denials, and failures, and impossibility of attaining peace or certainty, He bids us think again. All evidence, all demonstration, that can be given He heaps up against us, to show us we are wrong, and must think otherwise of Him, and of His universe, and of ourselves. Plainly He says to us in language of unmistakeable experience, 'Awake, thou that sleepest.'

CHAPTER VII.

OF NEGATION.

Wherefore do ye labour for that which is not bread?

Love is not-love
When it is mingled with respects that stand
Aloof from th' entire point. *King Lear.*

THE universe is more than it is to man, and to think rightly of it man must remember this, and consider his defective being. The simplicity of the conception constitutes its sole difficulty. The change in our thought is, in truth, so slight, so amply prepared for, so little requiring subtlety of apprehension, that it is difficult not to think it greater than it is. Not a disc in the heavens and a moon besides, but the moon perceived as a disc; not a physical world and another spiritual, but the spiritual perceived as physical: too little to us, its life and being wanting, therefore inert, therefore transient and unreal. Not because there IS that which is inert and transient, but because that which IS is not felt by us as it truly is. Simply we do not know that which IS, the true essential being; therefore, when that which seems the reality to us is treated as the true reality, inevitably it betrays itself as a delusion and a snare. Only a speculative opinion is given up in affirming nature to be truly spiritual. It is not denied that it is felt as physical by us. This is the proof of man's want of life. Feeling, acting, working, perceiving, remain the same; only our belief respecting their cause is changed, raised from less to more, from difficult to simple, from inadequate to worthy. There is another difficulty indeed, but that is not intellectual; it arises from the moral bearings which such a change in our thought has upon ourselves. Hence so prolonged a labour has been necessary to free us from an assumption so soon and so easily proved to be untenable. The inseparable connexion of these questions with man's conceptions of himself, his life, his relation to God, accounts for all the course of his thoughts in respect to them; and especially for the hold he has maintained on the accuracy of his impressions, and the existence of that which he feels to be. Only when a firmer basis is given for his belief, can he let go his natural conviction; only when a true reality is shown for his rest and confidence, can he quit his grasp of the false and treacherous reality, to which in his ignorance he clings, with trembling resolution to believe, and fierce wrath against all who, either in levity or in seriousness, point out to him how vain his confidence is, upon what a quicksand he is building.

Just so protracted, just so apparently hopeless, with good grounds on each side, should the strife have been. Looking back from the vantage ground which the recognition of man's deadness gives, we see that the progress to it must have been such as it has been. The defectiveness in man, which creates the problem, necessitates all the difficulty which has attended its solution; necessitates our own embarrassment, and doubt, and imperfect satisfaction; the strife between reason and feeling in ourselves; the inability to receive the assured conviction that this, which is to us so poor and mean and evil, can truly be, if we could feel it rightly, so good, so perfect, so glorious. The forms which appear, the things which are not eternal, and that which is true of these forms, these are so real to us, we can hardly feel that they are not the true realities, that they are but phenomena; the sole fact being unutterably above them, utterly unlike, and only to be learnt from them by most careful remembrance how unlike that which we feel must be from that which is. We can hardly credit how wrong man must be, to feel so wrongly; how dead, to find the universe so dead, if it be not truly so. More easy is it to us to believe that the evil and defect are not in man; it does less violence to our natural persuasions to attribute them to nature, even though they fall upon her Author.

Thus the sensuous feeling, and the foregone conclusion of the mind, struggle against evidence. But they ought to struggle. The doubt adds certainty to the proof; the difficulty testifies against itself. Man does feel that to be which cannot be, and all that is thereby proved of him is true. It is only needful that we should admit the evidence of all our faculties; unite them, and not deny or do violence to any. Consciousness testifies the reality of physical things to our feeling (it cannot testify to more), reason testifies that they are not truly real. Putting together the evidence of both, giving them both their full weight, the conclusion is evident: that which is not truly real is felt as real by man. Is not this simply a want of being, a want of life on the part of man? And what so natural, what difficulty in believing it? Why should we repudiate the testimony of either faculty, why coerce any part of our nature, when their united testimony agrees so well, and issues in a result so true?

In respect to existence, we have had such a perplexity as a child might find in dealing with the *minus* quantity in mathematics. We have been embarrassed in dealing with the negative elements in thought. The inaccuracy of our mode of thinking is strikingly shown, indeed, in the feeling that we have respecting the word *negation*, or the idea of negation, when it is definitely put before us, as if it were something strange, far-fetched, or at least abstruse and technical; while in fact no conception is more familiar, or more constantly in our mouths. Almost half the words in every language express negations, and negative forms of speech are of constant recurrence. That we, dealing thus abundantly with negations, should be startled as if it were some fanciful theorizing, at hearing negation spoken of as a matter to be considerately regarded, is a most instructive thing, implying some great inconsistency in our thoughts. And there is an inconsistency not less evident than harmful. We regard negations as facts or existences, and that for no other reason than that they affect us, or seem to affect us, as if they were so. We are, in respect to our conception of existence, and our attempts to solve the problems connected with it, as a boy would be in respect to an equation, who took all the *minuses* for *pluses*. For this reason and no other, we say the problems cannot be solved. If we will not recognise the negative character of some of the elements with which we have to deal, certainly we never shall solve them. Thought exists by virtue of negative conceptions, even as algebra exists by minus quantities. To ignore them is to deprive it of its power as a means of discovery. This is remarkably shown in that field of physical research in which such triumphs have been won. Throughout science the negative is recognised as of equal scope and importance with the positive. Life and death, light and darkness, heat and cold, the presence and the absence of whatever element or power may be in question, are equally regarded. The region of philosophy has differed, hitherto, from that of science, in the want of a regard to negative elements. The observer of physical forces pays constant heed to negation, or absence, of force; the inquirer into being has hitherto had little regard to negation, or absence, of being. Hence one chief cause of the failure of philosophy. It is revivified by a regard to negative conceptions; it is made triumphant by a recognition that the perceived negation must be referred to man. For, in truth, some speculators have observed, and justly insisted upon, the presence of negative elements in that which is perceived: but this can be no solution. It is a statement of the problem, not an answer to it, and a statement naturally felt to be repulsive. The recognition of negation perceived without us can be nothing but a preparation for knowledge of the defect within.

Evidently, the assertion of a negation in respect to man is but the transference to philosophical, or technical, language of the statement of his want of life. The two expressions differ in form; the former being abstract and indefinite, the latter more practical and explicit. But we cannot fail to see that the Scriptures do, in the plainest terms, assert a negation in respect to man. The novelty of this form of words should not blind us to its meaning. It is an expression proper to thought, as the affirmation that man has not life is proper to religion. Nor should its abstract form make us overlook the simplicity of the proposition; or prevent our seeing that it is but another mode of saying that which we continually say, in words more familiar but not at all more simple. We say, man is imperfect, of inadequate faculties, in a low inferior state, clogged and limited in his powers by his physical condition, subject more or less to a strange power of evil within him. All these things are indirect, imperfect modes of saying, according to the language of philosophy, that there is a negation or defect in respect to man, and that he is such as he is by virtue of that defect.

It is well worth while to familiarize our minds with the idea of negation, and to see what it truly means, since we continually have recourse to it in our thoughts. Why is it necessary to our conceptions? This question is not hard to answer. The idea of negation is necessary to the intellect, for the very reason that by intellect the true fact of being cannot be known. The necessity of the conception of negation denotes the fact that our conception cannot grasp that which truly is. Hence, in part, the natural repugnance to the idea, and the ludicrous aspect which it presents on its first suggestion as a distinct subject of thought. That we must recognise negation shows, at once, that our thought deals not with that which IS, but with that which only appears. Negation may be the phenomenon, may be perceived, may exist relatively to us, but it cannot truly BE. Doubtless it is for this reason, that the recognition of the negative elements in thought, as being negative, has been so tardy. For to know that that which is to our thought, which alone we can conceive, is not truly that which is, also involves the recognition of that death of man, which it has been so hard for him to learn: involves that he feels as reality that which is not real.

That the intellect, therefore, in attempting to conceive being, is compelled also to admit negation, means simply that the intellect cannot truly conceive being. It is compelled to deny that which it asserts. To the intellect, being must appear under a twofold form, of being and negation, for true being is more than intellect can grasp. Or it may be thus expressed: Thought demands opposites, can exist and operate only by contrasted ideas; therefore if thought deal with the idea of being it must also entertain the opposite idea, which is that of not-being, or negation. Not because there is any negation, or can be, but because of the necessary laws of thought, as demanding opposites. So far the reason of our necessity for conceiving negation is not difficult to understand. The idea appears strange or foolish to us only because we have not reasonably considered it; have not reflected on our own words and thoughts.

Negation is necessarily relative: it cannot EXIST. The existence of negation is a contradiction; but there may be negation relatively to any particular thing, or mode of being. Negation must enter as an element into all relative knowledge. It pertains therefore to thought, the scope of which lies in that which is relative. Negations are perceived; they are felt by us as existing, and as producing effects. This is easily understood. Negations appear to act by virtue of the operation of that of which they are the negation. Cold produces effects, and seems to be a power in nature, not because it is anything, but because it is the absence (or negation) of heat. Darkness, also, produces effects in connexion with light. In themselves darkness and cold are nothing; but as the absence of that which operates, they also appear to be operative. It is worth noting how large a part of human activity is caused by negations. Absence of heat and absence of light, what exertions, what widely extended operations, to remedy or to employ, do they institute among men.

To say there is negation in respect to man involves no difficulty in the conception, as if a negation existed, or anything were absolutely wanting. It implies, only, that in regard to man, taking the true being of humanity as the standard, there is defect. The case may be illustrated very simply. Nature includes many imperfect things, such as undeveloped plants and animals. The perfection of nature is thereby in no way impaired; it is perfect in including these. But if the fully developed, or perfect, animal or plant be taken as the standard, then the idea of a negation is introduced. There is negation in respect to it. Or again, when an organic body dies, no defect is thereby introduced into nature; other life perfectly takes the place of the life that has been lost. No negation EXISTS, but there is a negation in respect to the body that has ceased to live. Regarding the case from that point of view, the negation is perceived.

The Life of the universe is perfect and unmarred; there is no defect in it. But when regard is had to man, negation is perceived: his life is defective. To think aright of him, to understand his condition, regard must be had to a state of being which is not fulfilled in him, to an existence in which there is more. This should not be hard for us to conceive. So we think of a corpse with reference to a living frame; so we look upon the immature animal with reference to maturity. We cannot understand, we do not attempt to understand, the structure of a developing organism, without a constant reference to its perfect state. To conceive rightly of man as he is now, we must ever keep our eye fixed on another condition of his being; his true being as man, from which this life differs by defect, and only for the sake of which this life is, or could be. Because man must LIVE, therefore, and therefore only, could he pass through this living death.

By defect, or negation, therefore, man is physical, and perceives the world as physical. That which has no true existence is felt as the reality of his life, and compels him to infer a material or inert substratum, until he knows why it is that he feels that to be which is not. Availing ourselves of the artificial language invented for other subjects of thought, that which constitutes physicalness may be compared to the addition of a minus quantity, that is, it is a loss. Our conception of the physical, as something added to the spiritual, is as if it were supposed in mathematics that the addition of a minus were a real addition. The simplicity of this idea, and its conformity to the feelings which we truly entertain upon the subject, is shown by the thoughts of religious men respecting the death of the body. They feel it to be a gain, the liberation from an encumbrance, the passing to a state of more perfect being. To be freed from the physical is, to their apprehension, to be freed from a defect. To say that we are physical by loss and want, therefore, is but to interpret the language of the heart into that of the intellect, to bring our mode of thinking into unison with our truer and more manful mode of feeling. Nor is this argument less applicable to those who hold that there is no consciousness without a body, and that the resurrection is the commencement of the future life. For not less do they hold that the future body will differ from the present in not being physical. It will have been freed from a loss, have lost a defect. 'Not that we would be unclothed,' says St. Paul, 'but clothed upon, that mortality might be swallowed up of life.' Thus our thoughts are simplified and harmonized, confusion and difficulty are removed. Such a relief is given us as would come from recognising, for the first time, that to add a minus was to sūbtract. The physical comes by adding a minus.*

Thus a recognition of the negative character of that which is negative in our perceptions, simple as it is, effects an enormous gain for us, places us in an entirely new attitude to the problem of existence, and, above all, enables us to embrace and unite in mutual helpfulness, discordant and hostile views. It disarms scepticism of its sting; emancipates devotion from its bigotry. It leads us to give up self, to feel that we ourselves are no standard; it frees the intellect from bondage, the heart from cruel constraint. We are no more compelled to hold that to be real which

* Mathematics is not peculiar in dealing thus with minus quantities. It is the type and exemplar of all thought; doing consciously, and with understanding, what is done unconsciously, or ought to be done, in other departments of intellect. In great part, its superior certainty and power depend on the clear recognition of the part played by the negative element. Other branches of thought have in this respect to conform themselves to its laws. The example of mathematics proves, at least, that the conception and use of a negative idea, as a means of interpreting nature, is natural, reasonable, and conformable to experience, and ought not to be embarrassing.

we feel as real, and are therein set infinitely free. We know that things are not what they seem, not what they are to us. Man's universe gives place to God's. For this that we feel to be cannot be the very fact of existence, conceive it, or conceive as added to it, what we may. It is not enough. In asserting it we assert negation. This is the simple substitute for all speculations respecting matter, and properties, and essence:—Man wants life; that which is to his feeling and his thought is not that which IS.

We are as a blind man, who may be said to be conscious of a negation by his defect of sight; we are conscious of a negation, conscious of physicalness, by our defect of being. And as no knowledge can make a blind man otherwise than conscious of a negation, nothing but the removal of his defect, though he understand ever so well that the world is not as it is to him; so we may understand perfectly why we are conscious of negation, and that the true world is not as it is to us, while our consciousness remains the same. Yet it is of the greatest moment to a blind man to know that the world is not dark, since only so can he be brought into right relations with his fellow-men, know his true position, or, above all, embrace the means of cure. And of the utmost moment is it to us to know that the world is not physical. Only so can we assert our right position, or recognise our true relations.

And, further: a man, born blind, does not consciously feel that there is any negation in his perception. He is not naturally aware of any defect on his part. Even so, we do not feel that in our perception of the world there is any negation; we do not naturally recognise in ourselves any defect. To us it seems that the physical is emphatically the real. But we learn the defect in our perception, as the blind man does, by the evils, the mistakes, the failures, to which it subjects us; by the disproportion we find between our instincts, our desires, our native endowments, and the results we can attain. We do not perceive aright, we come to mischiefs and injuries unforeseen; fear takes possession of us, in the midst of day we grope as in the night. We cannot act aright, nor adapt ourselves truly to the world in which we are; for we do not know it rightly, it is more than it is to us.

BOOK III.

OF RELIGION.

Then I saw that God hath a larger mouth to speak with than I had a heart to conceive with.—JOHN BUNYAN: *Grace Abounding.*

The order of subjects necessitates, here, a word of explanation. The views advocated in the following chapters are not an offshoot of, nor a deduction from, the theoretical conceptions which have preceded. They have been drawn immediately from the study of Scripture, and from them the philosophical ideas have largely derived their origin. The arrangement which puts the philosophical views before the scriptural has been adopted for the purposes of the book, but that order does not represent either their historical or theoretical dependence.

CHAPTER I.

OF DEATH.

Nor is it at all incredible that a book which has been so long in the possession of mankind should contain many truths as yet undiscovered. For all the same phenomena, and the same faculty of observation, from which such great discoveries in natural knowledge have been made in the present and the last age, were equally in possession of mankind several thousand years before.—BUTLER: *Analogy of Religion.*

MAN is in no other sense prejudiced than as he clings to that which he cannot feel himself justified in resigning. He is not unwilling to advance, but he is fearful. His very timidity, and consciousness of his liability to err, drive him to assume positions which only the most perfect self-confidence could justify; for there is no rashness like that of fear. We are so bound to that to which we have been accustomed, because to us the unknown is full of vague terrors. To exchange that which has been felt as certainty, for that which seems uncertain, because untried, is painful to us. We fear disaster wherever we cannot see. The instinct which makes the stoutest heart shrink from darkness, and peoples it with phantoms, is equally strong in the intellectual world. On the accustomed principles certain results can be secured, and we wish to rest. We are content not to be wiser if we can but feel sure. But God will not let us rest. He has other work for us to do. Above all He will cure us of our mistrust. For the secret of this misgiving is that man has not faith in God. The evil of his nature shows itself in fear. He that is conscious of wrong must be afraid. Adam, when his conscience had awakened, hid himself at the voice of God; so do his children hide themselves at the voice of truth. It was a just fear in Eden, it is a just fear now; but it issues now, as then, in a foolish deed. At the feet of God the shrinking conscience must regain her peace; the timid intellect renew her daring, bowing herself to truth. We do right to fear, we do right to come to that of which we are afraid, that the cause of fear may be taken away.

Men adapt their moral and religious convictions to their intellectual conceptions. That which the conscience and the affections demand will always secure for itself a place in the human belief, but these moral elements entwine themselves with others purely intellectual, and the whole are built together into a connected and more or less completed system. Thus it comes to pass that the intellectual convictions, being held authoritative, become the basis, as it were, of the moral and religious beliefs, and determine their form and expression. When, therefore, our intellectual opinions are in any considerable degree affected by the results of inquiry, it necessarily happens that our moral beliefs appear also to be implicated, and much embarrassment is caused in this way. It is hard to remember the essential independence of two things which have been thus closely united; and all the power of the moral nature is often enlisted on behalf of merely intellectual opinions, which are in themselves most antagonistic to religious principles, but to which those principles have been, more or less laboriously, reconciled. Often, also, the more opposed to the spirit of religion a particular opinion is, the more intimately it appears to be involved with our religious convictions; for the very reason that a larger amount of toil and thought has been bestowed upon the task of bringing them into even an appearance of agreement.

The writers of the New Testament declare men to be dead. They speak of men as not having life, and tell of a life to be given them. If, therefore, our thoughts were truly conformed to the New Testament, how could it seem a strange thing to us that this state of man should be found a state of death; how should its very words, reaffirmed by science, excite our surprise? Would it not have appeared to us a natural result of the study of nature to prove man dead? Might we not, if we had truly accepted the words of Scripture, have anticipated that it should be so? for if man be rightly called dead, should not that condition have affected his experience, and ought not a discovery of that fact to be the issue of his labours to ascertain his true relations to the universe? Why does it seem a thing incredible to us that man should be really, actually dead; dead in such a sense as truly to affect his being, and to determine his whole state? Why have we been using words which affirm him dead in our religious speech, and feel startled at finding them proved true in another sphere of inquiry?

Do we say that man is 'spiritually' dead? That is the very thing affirmed by science. Spiritual death is actual death; death in respect to true life and being: the death which constitutes the world a dead world to us. Man is dead to the spiritual, dead to the eternal, dead to that which IS; so that mere passing forms are the realities to him. Science reveals to us a result of man's being spiritually dead; shows that death to be a profounder, more real thing, more truly worthy the name of death, than we had thought it. That death causes our life to be not truly life: a life to that which IS not.

When we see that there is a deadness in man, scales fall from our eyes in reading the Bible; our thoughts are in harmony with it. For one chief part of the wonder of that book lies in this: that whereas we have taken it for granted that man has his life, the men who wrote those pages knew that he was dead. They are saying what every man in his soul affirms to be true. Those words are the fulfilment of that which all men long for, which all men recognise. But they are truer than our thoughts: if we would do them justice, we must take them as they are, not bending them to our conceptions. The affirmation of the New Testament is that men are dead, and that they are to be made alive through Christ. But we have been compelled to make the Bible affirm man's life, and have, therefore, given to its words whatever meaning we could best, and most reverently, put upon them consistently therewith. For by reverent and loving hands has this violence been done. Not through fault, but unavoidably. How could it have been otherwise? While science seemed to be demonstrating deadness throughout all nature, how could man's deadness be maintained? Was there to be no life in the universe at all? For to our ignorance the uniformity of nature seems a dead necessity, and we cleave therefore to the life and spirituality of man as the only basis on which a religious faith can rest. We are compelled to deny that man can be inert, compelled to assert for him freedom, compelled to take his state as the type and evidence of spiritual being. There has been no choice before us. We have been compelled to interpret the words of the New Testament conformably, for we could see otherwise no possibility of religion at all.

But can words more plainly affirm that man is not spiritual, that he has not life? Must we not have been laid under constraint, subjected to a perverting force, in interpreting the Scriptures? Does not the recognition of death as the state of man, come from what source it may, set free the Bible from conceptions alien to its spirit? May we not ask ourselves, whether our religion, though based honestly and most earnestly upon the words of Scripture, have not involved, in spite of ourselves, a bending of those words to suit our imperfect knowledge?

Nothing can be more striking than the simple way in which the deadness of man is laid down in the New Testament. It seems almost to be assumed, as if it were a thing known and evident, not needing to be proved or made matter of special demonstration. As is the existence of God to the Old Testament, so is the deadness of man to the New; the fact central to the whole, the postulate, as it were, on which the entire volume rests. May not a reason be, that the death of man is a central fact of the Old Testament also; that man died in Adam? became such as he is through that transgression? Therefore, when the New Testament writers take up the history and tell of life bestowed, of a true spiritual life bestowed on man, of necessity they speak as they do. It is not theirs to prove the death, that is the known, the evident fact, only theirs to reveal the deliverance from death. The burden of the New Testament is that man is to be made alive; he is to be saved from death.

If this be the Gospel, what a glory follows! What light and joy break in upon this dark and miserable world! We may almost begin to see it as God sees it, and understand that our ignorance alone has clothed it in such appalling gloom. If this were man's life, truly it were a dark, a fearful, a mysterious world; a world to fill with despair the most trustful heart, and tax too much the strongest faith. But if it be man's death, all is clear. That which cannot, may not, must not, be man's life, may be his death. How should his death not be even such? What should death bring but sin, and folly, and delusion, and agony, and vain grasping at shadows, and sickness, and remorse? What but this world should be, could be, the fruit of the death of man? Knowing the death and the redemption, the very spirit of prophets and apostles comes into our hearts, and their words become so simple and so true. Man is dead, therefore we are as we are; and God has saved us, and will make us new, yea, give us life in Christ.

For the impossibility we have found in recognising that Christ is the Saviour of the WORLD, in believing that He will draw ALL men unto Him, arises from our belief that this state is the life of man; from our not having been able to see that the New Testament calls it death. If we can alter our point of view here, all the else insuperable difficulties in the absolute redemption of the world are gone. Clear and consistent, satisfying all demands of conscience, and heart, and intellect, stands before us God's scheme and solution of the world: man, from death, is being made alive.

Surely we should let every book explain itself, and be judged by its own words. The Bible may be rejected as a guide; every one must judge for himself of its value: but to interpret it against itself is to do a grievous wrong. To ignore, when it speaks of death, that it has defined death, and expressly stated what it is; that it speaks of the present state of man as a dead state; is to deal it hard measure. How can we be surprised that, dealing thus with its language, we should be conducted to results which appal our hearts, and baffle our thoughts, and clothe in tenfold mystery the already too great mystery of life; that although we call it the Book of God, it remits still to the future those great questions of His love and justice, which it is the very life of our souls to know. We do too great a wrong, and reap too severe a punishment. For blackness and darkness close around our souls, and our hearts groan with an anguish that will not be subdued, that faith itself cannot calm, nor the very love of Christ cheer with one gleam of hope.*

But when we are seeking to understand the Bible, what does it matter what *we* think is life, what *we* think death must be? The sole question is, what does that book speak of as life, what does *it* term death? Using its words consistently with themselves, nothing can exceed the simplicity of its statements. For are they not summed up in this: that Christ has died for men that they may be saved from death, and that believing in Him they shall have life? What affirmation can be plainer, if we remember that the same testimony has affirmed that men are dead? Christ has died

* See what that sincere and earnest man, Henry Rogers, says:—'For my part, I fancy, I should not grieve if the whole race of mankind died in its fourth year. As far as *we* can see, I do not know that it would be a thing much to be lamented.'—*Greyson's Letters*, Second Ed. p. 22.

to save men from the death in which they are. The same men that proclaim the sacrifice of Christ, to obtain life for man, proclaim his present death. The two statements are integral portions of one whole. To separate them is to distort and to destroy.

We have been regarding the death from which Christ saves as temporal, as a thing which may be postponed. But it is eternal; it has relation to man's actual being, not to changing circumstances. All our embarrassment has arisen from our not having been able to perceive that this is death, that man is now and truly dead; from bending all the words which declare it into another meaning. Some have said, the death is future, man is condemned to die, in danger of dying: some, it is a figure: some, it is a death indeed, in a spiritual sense, but different from that true death of misery which is to ensue hereafter. But all these opinions have had one basis; inability to believe that this state, which men like so well, could be the death spoken of in such terms of awe. From the death of man has come this thought; the saddest fruit, the most convincing proof, of the very death that is denied. For what is it we are saying, but even this, that mere wickedness, mere self-indulgence, merely being alienated from God, is not worthy to be called death, unless there be misery conjoined with it; that suffering is more to be feared than sinning? In that speaks the death of man, it needs no more words to prove it; that is death which makes man fear suffering more than sinning.

CHAPTER II.

OF LIFE.

We know that we have passed from death to life.

FROM the state in which man is, Christ died to save him. His life he gives for man who has not life. So we are made to know God in the true sense of knowing, and in that knowledge have our part in the life eternal.

All the difficulties which have rendered the nature of the eternal life bestowed by Christ a matter of dispute, resolve themselves when it is remembered that man is dead. For the affirmation of the New Testament is, that in Christ is given to men a life which makes them alive from death. Therefore this life is the opposite of the death in which they are. If that be eternal life, then is this eternal death. So that by the death which we know, we may know also what the life must be. But as we cannot know this life by sense, so neither can we know it intellectually. We cannot think it. A chief part of all the difficulty that has beset religious questions arises from our resolution to conceive the eternal. It cannot be conceived. It is to be known spiritually, actually; it will not be put into our thought. Having learnt that by our intellect we can know no veritable fact at all, but only the appearances of things, how should we suppose that by the intellect we should know the eternal? We have made ourselves the standard, and projecting our own deadness into an endless future, have called that eternal life; but God's thoughts are not as our thoughts. Eternal life is that true life by want of which man is such as he is. It is spiritual, actual life; a life known within the soul, but not to be conceived in thought. Whatever is so conceived becomes in that very process no more eternal.

The first necessity for a right attitude towards the eternal is, that we should abandon the supposition that our intellects can conceive it. The eternal is that unknown fact, of which all the things that the intellect deals with testify, but which is not in them; the want of which in us prevents our knowing that hidden fact. When science and philosophy unite in testifying that the essence of things is not by them to be discovered, they do but reaffirm St. Paul's declaration, that spiritual things must be spiritually discerned. The eternal is known in living. It is the fact which the intellect has proved and confessed its inability to grasp. 'Eye hath not seen nor ear heard, neither hath entered into the heart of man the things that God hath prepared for them that love Him; but He hath revealed them to us by His Spirit.'

That we do not know the eternal is our death. Understanding that this state of man is his death, it is no more a difficulty that we cannot think the eternal, nor conceive it, and must put aside that which it is our natural tendency to suppose. If man be dead, it should be so. The living knows the eternal; it were a contradiction that the dead should know it. Let the intellect take its place, as dealing not with the very fact of being but with phenomena, which is now no more a religious dogma only, but the accepted result of physical and metaphysical research, and the meaning of the eternal ceases to be difficult, ceases to be fraught with painfulness, either to the intellect on the one hand, or to the heart upon the other. The eternal ought to be, as it is, inconceivable by thought, else it could not be true being; else must it also be a passing, empty show, like earthly things. To know the eternal is to Live.

Once let it be seen that there is a deadness in man, and all is simple. He is no standard; his necessary conceptions have no authority, are necessarily wrong. That which truly IS cannot be according to his thoughts, but he has to be made different, to be raised to a truer state of being and of feeling. Yet it is not difficult for us to know that there must be a state of being differently related to time from ours; a life to which the phenomenal things are not the realities as they are to us. We grow old, the lapse of time affects and alters us; our being is in time, and is determined by its course. But not so is God. He is not older. Time and things in time are not to Him as they are to us; not the realities of His existence. They are to Him as they truly are. To be in time is for mere phenomena to be our realities. God is as a rock beside which flows a stream; we are as a straw which it bears along. That is the eternal life which God possesses; of that man is to partake.

'This is life eternal to know Thee.' Eternal life is that which Christ had given to His disciples, because in seeing Him they had seen and known the Father also. In believing in Christ we pass from death unto life. But it may be said: believing in Christ does not make us different; we remain just as we were, except in the feeling of our hearts.

That is true. The individual life does not remove the deadness of man. That deadness, as it does not arise from a condition affecting the individual alone, so it cannot be removed by an individual change. Therefore, the man who has received eternal life from Christ is described thus by St. Paul: 'I delight in the law of God after the inward man, but I see another law in my members, warring against the law of my mind, and bringing me into captivity to the law of sin, which is in my members. We wait for the adoption, the redemption of our body.' In him there is life struggling with death; a life that is given to him by Christ, a death that he partakes with humanity. The perfect redemption of the individual is in the redemption of man. When man is saved, then there is no more death. Death is destroyed. God is all in all.

For eternal life is not everlasting happiness. Salvation is not being saved from misery. These are blessings which man's heart longs for infinitely more. In all his ignorance and wickedness, man is not sunk so low as not to feel that what he wants is something better than happiness, that the curse under which he groans is not suffering. He knows not what it is, nor how to utter it. His prayers are inarticulate, his toils a weary, undirected strife. But in his inmost soul he cries out to be delivered from himself, to be saved from the fatal spell that is upon him, whereby he must seek his own pleasure, to gratify and to exalt himself; to be saved from passion, from the inward gnawing death that leads him into all evil, itself the greatest of all evils. He wants life, to BE, to act, to be no more a slave.

Eternal life is given us in Love. God's own life put within man's breast. When man is made alive, we shall no more be compelled to pursue our own happiness, to seek for self-satisfaction. Love shall be made perfect in us. Our life shall be like God's, one with His, who lives in giving only. And while the death yet cleaves to man, still we have the life. A new being is within us; a life, a knowledge, a relation, that we had not before. For we know God: know Him truly as He is, the infinite, sacrificing Love. Our whole thought of life and good is altered; in giving up ourselves is all our glory, all our hope and wish. The whole fact of the universe is altered to us, for we know its life, and source, and centre.

We are delivered from the death that bound us. It is no more, as it was before, necessary to us to have concern for ourselves: by knowing God we have been made free. For knowing HIM, first we trust Him perfectly, and feel no more the need of caring for ourselves; and next, we loathe and detest ourselves that we are so unlike Him. All our heart and soul are changed. To have regard to self is hateful to us, for we know that that is death. All that is good or lovely in our eyes is in utter sacrifice. Suffering and loss have terrors for us no more. Our joy is to be one with God in giving: we want only the perfect deliverance from death; we rejoice with joy unspeakable, knowing that God is making that deliverance complete in giving life to all men, we also being fellow-workers herein with him. Our life is eternal. We know that passing things are not the fact with which we have to do. We look for the crown of life in heaven; the crown of life in love perfected, in sacrifice made complete.

CHAPTER III.

OF DAMNATION.

IMOGEN.—I draw the sword myself. Take it and hit
The innocent mansion of my love, my heart.
PISANIO.— Hence, vile instrument;
Thou shalt not damn my hand. *Cymbeline.*

He that doubteth is damned if he eat.

If thy hand offend thee, cut it off: it is better for thee to enter into life maimed, than having two hands to go into hell, into the fire unquenchable; where their worm dieth not, and their fire is not quenched.

He that believeth not shall be damned.

OUR fatal habit of putting the eternal at a distance from us perplexes all our thoughts of Scripture. Therefore it is that we hardly dare to speak the word damnation, that to utter it seems like sacrilege: we have put it so far away. But that which the writers meant was not a thing they were afraid to speak of. They had not banished it into the future. The damnation of which they spoke was a thing that IS, an eternal thing, the true and actual death of man. Men are damned in sinning. Why is it that when the Bible speaks of death and of damnation as present things, we reduce them to so small a matter; but when it uses the same words with a reference to the future, immediately we fill them with a meaning the most awful we can conceive? Why do we make this distortion of its language; why put its words thus upon the rack, and cramp or stretch their meaning according to a rule of tenses? Do we not deal thus with the Bible, because this state of sinfulness is pleasant? We cannot believe that this is really the damnation, because men like it. It never occurred to us that to like to be wicked could be to be damned. That was not bad enough.

Here we behold ourselves. We have taxed our thoughts to find the worst thing that could befall, and have invented suffering. Of all the many sins we must confess to God, is not this the head and chief? It is without excuse; for it is a violence not only to the light of nature, but to the plainest use of the very words on which the meaning has been forced. The fatal virus of the disease has turned the very medicine into poison. For to what end is the Bible written, but to make us know and feel the awfulness of sin, to make us afraid of sinning, to rouse the capabilities of our nature to a knowledge of what it is to be opposed to God, to instil into us a fear of wrong as wrong, to give us that new feeling which should make our hearts respond to words which describe a sinful state as the chief of evils? Why should not sin be treated as the most awful of things? Is it not so? Do not all men, as they approach to God, more and more feel it so? How, then, should God not speak so?

Sinning is damnation; self-indulgence is to be cast into hell. These are the most fearful terrors, the chief of evils, in the sight of God. Let the words be read and tested. The difference between that thought and ours is the difference between life and death. Does not that which we most like or dread depend on what we are? Self-indulgence is hell, the worm that dieth not, and the fire that is not quenched, unbridled passions. It is better maimed to enter into LIFE, than having two hands to go into hell. But what life is we know: it is the opposite to self-indulgence, the being one with God; and unchecked passions are an unquenchable fire, a consuming flame that is never slaked, that burns more fiercely in the soul with each attempt to quench it. No words can be more simple than are these; it is their truth that makes them difficult. That the worm dieth not, and the fire is not quenched, proves them not physical. Physicalness is excluded by denial of its characteristic property of ceasing. A worm and fire not physical, what should they be but the devouring and consuming passions, consuming ever, yet leaving unconsumed; or how should a true-seeing man speak of them otherwise? 'Enter into life with loss of all things, but be not self-indulgent, be not cast into hell.'

The message of the New Testament to men is that they are damned, and they know that it is true. They do not fear a future damnation that is not like the present. Men know that they are damned: that the burning passions in their hearts are never quenched, nor can be quenched; that one desire sated, another takes its place; that the imperious appetites are their plague and torment. To preach another damnation to them than this, is to pervert the Scripture, to array against ourselves the truest feelings of humanity. This is eternal death, the lake that burneth with fire and brimstone. It is here and now; to see it we need only that our eyes be opened, we need only that the life that was in the men who wrote should be in us, who read. For do we ask what hell-fire is? God has answered us. 'If thine enemy hunger feed him, and if he thirst give him drink; for in so doing thou shalt heap coals of fire on his head.' Here is shown us hell: God overcoming evil with good, consuming man by bounty. For what is it fills our hearts with passions, and burns us up with the fire of insatiable desires, but God's own gifts, the charms of nature, the good things with which He crowns our life? God's gifts kindle the fire within us, His bounties are our torment. So he casts us into hell, surrounding us with good; for love is fire. To be loved by a man whom we treat as an enemy, is to have coals of fire heaped upon our head. To be loved as God loves us, we being such as we are, is to be cast into a lake of fire.

God saves us from hell, saves us from damnation; saves us through believing. He who believeth not shall be damned; must be, certainly, inevitably, will be. He who believes not on Christ, who does not know God in Him, know Him to be such as Christ reveals Him, will be damned.* He will be wicked; there is no escape from sin, and the dominion of self, but by faith in Christ, for, save through Him, God cannot be known. There is no other name given among men whereby we must be saved. If God be not known as Christ reveals Him, nothing else can avail to extricate man from death.

The men who wrote about damnation in the Scriptures saw things rightly; they had true perceptions, feelings justly attuned to the reality. They were living men. To them, to sin was to be

* In the original language of the Bible there is no distinction for shall and will. Either word may be used as seems most expressive of the sense. Perhaps the beauty is impaired sometimes in our translation by the use of shall where will might be better.

damned; they put sinning above and before all other things as the great and chief calamity; the fact that there was a power, a life, that could save men from sinning, filled their hearts with wonder, and made their lips overflow with words that cannot be forgotten. That fact made them beside themselves; dwarfed all things else in their regard; Christ crucified was all to them. By His power on themselves they knew that He could give life to the dead, for He had quickened them, and raised them up, and made them dwell in heaven; so that the life they lived was not their own but God's in them. How could they doubt that He should save others also? complete the work He had begun, and take away the sin of the world? For the heart of man responds to their words, what they have affirmed is true. Sinning is worse than suffering; we know and mean it in spite of our own words. To be damned is not to be miserable, but to be bad. The love of Christ, the sight of God as He truly is, must have power to save all men from sin. Christ must draw all men to Him.

But although damnation is not suffering, it is not therefore to be inferred that suffering is not threatened as punishment for sin. It is so threatened. All sin brings punishment: God will render to every man according to his works. But it is from damnation, and from death, that Christ saves us. His work has relation to the eternal. Our view here has been obscured, as everywhere, by treating the eternal as future. Throughout the Scripture, Christ is spoken of as saving us from sin, from corruption, from vain conversation, from this evil world, never from pain; from that which is the worst thing, not from that which we most fear. Our familiarity with the latter idea renders us unconscious what we read. The death that is our present actual state, our condition in relation to the eternal, is that evil, fearful thing from which Christ has died to redeem us. In this He makes God at once just, and the justifier of him that believes. Our thoughts being other than those of the Bible, we have with great effort adapted its words to meet them; we have transferred them from the eternal and spiritual that IS, to the suffering or happiness that we look forward to. Our conceptions are so moulded to this latter idea, that it is difficult to look simply at the words of Scripture, and see how much more is in its declarations than in our thoughts. Christ died for the world, to save it from the curse of death under which it is; not a future death of misery, but an actual death of worse than misery; a death which involves our liking that which is evil.

We find it hard to believe that damnation can be a thing that men like. But does not what every being likes depend on what it is? Is corruption less corruption, in man's view, because worms like it? Is damnation less damnation, in God's view, because men like it? And God's view is simply the truth. Surely one object of a revelation must be to show us things from God's view of them, that is, as they truly are. Sin truly is damnation, though to us it is pleasure. That sin is pleasure to us, surely is the evil part of our condition.

Suppose there were a child who liked to eat dirt, should we not tell him that it was filthy? But how could he think that to be filthy which he liked? Would he not suppose that we referred to some consequences, to some future, which he would not like? When God warns us of damnation, present and future, we, liking it, think He is speaking of consequences. But the instructed child learns that to eat dirt is filthy, then he understands his teacher; man learns that sin is damnation, then he understands his Maker.

A sinful state is eternal death. It is death in relation to that absolute being which the eternal denotes. The application of the word translated eternal to an everlasting duration arises from our misapprehension of man's present state, from the false conception we entertain of all things, through our ignorance of man's want of life.* Men are now dead or damned eternally; a state from which eternal life raises them. An eternal state is one which relates, not to our conditions or circumstances, but to our very being. Christ gives eternal life, a life that makes us truly and absolutely living; not such life as the physical, which leaves us dead, being life only in seeming. So sin is true and absolute death, not like the physical death which does but appear to be so. In sin we have consciousness of death. 'Sin revived,' says St. Paul, 'and I died.' Doubtless man shall not like sin for ever. He must feel it differently, feel it the greatest misery. But how can that be to be damned? Surely that were rather to be saved.

---

* If the question be raised as to the grammatical meaning of the word *αἰώνιος*, is it not admitted that it was never used to express everlasting duration before the New Testament was written? Was it not expressly defined by Plato as not bearing that meaning, and applied by him to denote a kind of being not subject to conditions of duration? See the *Timæus*.

CHAPTER IV.

OF REDEMPTION.

Foolish and hurtful lusts, which drown men in destruction and perdition.

THE great fact of the New Testament is the redemption of the WORLD; the saving of all men through Christ. Perhaps no single statement is so often repeated, or asserted in so many different forms. It would be of no avail to enumerate the passages; for the most part they are well known. We have rather to ascertain why they have been interpreted to mean the salvation of part of the world, and the final loss of the rest. What necessity has acted on our minds to compel us to that conclusion, not less against the apparent meaning of those passages, than against our own deepest hopes and wishes?

The reasons have been of two kinds: in part the apparent meaning of other passages of Scripture, in part the evident fact that so many men do not believe, but die without participating in religion. Of these two elements, the latter is that which truly determines our opinion. For no passages can be plainer, or more emphatic, than those which seem to declare the absolute salvation of all men. No words can be more direct, or apparently decisive, than such as these: 'I, if I be lifted up, will draw all men unto me.' 'God sent not His son into the world to condemn the world, but that THE WORLD through Him might be saved.' If there be others which seem as directly to affirm that all are not saved, then it must be on other evidence that the interpretation is decided. One class of expressions will be taken as the standard, according to our general conceptions of the world and of man; and the others conformed to them.

Accordingly, the absolute salvation of man has been given up, because we could not otherwise understand that which we see in the world, and especially the fact that men die unsaved. That one circumstance outweighs in our estimate all other arguments. For we have not been able to conceive the world otherwise than as a probation for eternity. On this theory we interpret all those statements of Scripture which declare that all men are to be made partakers of life through Christ. It is well that we should see clearly what we do. For in truth, familiarity with the doctrine of a partial salvation of men has so moulded the thoughts of the greater part of the believers in the Bible, that they no longer see as they read the passages which affirm salvation of all, that they have even an apparent opposition to their idea. A certain meaning is so constantly associated with them, that men have almost ceased to be aware that they might have any other; their force and bearing are entirely lost. How many of those who think that some men go after death to a final and unredeemable ruin, are aware that there is in the Bible a single passage which opposes a difficulty to that opinion?

The doctrine of a probation for an everlasting destiny, and of the final misery and loss of a part of mankind, and happiness of another part, is man's natural supposition, from which the New Testament, revealing to us God, and life, and death, as they truly are, is the means of our deliverance. For a probation for eternity implies that this is man's life, that men are not dead; it implies that men are not in eternity, but that the eternal is a thing to come; it implies that men are not now damned, but only in danger of it. Therefore we feel so much difficulty in interpreting the New Testament, which says the opposite of all these things. For that men do feel this difficulty, the appeal may be made to their own consciousness. Simple it is, indeed, that Christ died to save us, and that believing in Him we have eternal life. Ever the conviction, in whatever ignorance held, that God sacrifices Himself for us, saves us from death, and makes us new creatures. But of the whole book, who will say that there is not great difficulty in reconciling it with our conceptions of the world, and with our conceptions of itself? Have we not taken our own view of man and of the world, instead of the view which it presents? Convinced by its evidences, have we not been trying to submit our thoughts to its words, while yet retaining a fundamental conception of ourselves, which those words emphatically set aside? For if we receive from the New Testament this thought, that men have not life but are dead, and that Christ gives life to the world, being the Saviour of all men from the death in which they are; that this salvation comes, as it can only come, through believing in Him; and that God wills that all men should believe; that not believing they must perish, or die, or be damned, because the only salvation from these things is by believing; that some men are elected, as we see, to believe, while others do not believe, and incur the penalties of not believing, but that their unbelief cannot make the faith of God of none effect, for He has said, 'I will draw all men unto me'; that

Christ, in the end, shall destroy death, and bring all things into subjection to Himself, so that God shall be all in all; if we keep steadfastly in view this fundamental fact, that Christ has borne man's death, and will make man alive, not partially but absolutely, and that all things else have their place subordinately and subserviently to this; then is not the difficulty gone? Not indeed that all the New Testament can be understood. God forbid that there should not be things in it which are beyond our present thoughts; passages which we must leave and say, I know not what this means. But the whole book is natural and simple. No statement in it embarrasses the intellect, or racks the heart with insoluble moral problems. The essential thing is that we should accept from the New Testament the statement that men are truly, actually dead. It refuses to be interpreted on the doctrine of man's life. Precisely there it is, that its thoughts are above our thoughts; so much above them, so much truer, so much in advance, that we could not understand it. The question of religion turns upon this point: Is this state, or is it not, the LIFE of man.

If we can see that man is truly dead, there is no more difficulty respecting the absolute salvation. For probation is thereby excluded, death and probation are incompatible. We naturally suppose a probation because we naturally suppose that we are possessors of true life, that freewill is freedom. It seems absurd to us, that this state of being should be the result of a loss and want of life. Necessarily this must have been our feeling, certainly not less if it be not true. Pride and great thoughts of self belong to death; more than all other things do they distinguish it. That man necessarily supposes his own state to be life, can afford no evidence whatever that it is so; none to weigh for a moment against the fact that he perceives deadness all around him, and finding no eternal in anything with which he has to do, thinks the eternal must be to come. Too plain it is that man is dead; the being to whom the eternal IS not.

It is true that we are practically under a probation: we are dealt with according to our works, are under a system of trial, of rewards and punishments. But this is not being under probation for eternity. These are means through which the work that is to be done in man is accomplished; means through which his death is removed, and he is brought to be no more the subject of rewards and punishments. They do not give the character of probation to his state, in relation to the eternal; and the express statements of Scripture exclude it. Only our own conceptions, overriding its declarations of man's death and absolute redemption, could have made us so interpret the statements which affirm the punishment of evil-doers. The latter in no way involves the former. Children are subjected to a system of trial, and are dealt with by rewards and punishments, but a school is not a state of probation.

Have not the words in the New Testament, which appear to affirm the final loss of part of mankind, received that meaning against their natural and necessary sense, because of our own preconceptions? Have not we ourselves put into the Scripture the teaching which we find there? These passages are of three kinds: those which speak of eternal punishment, those which speak of death as the result of sinning, and those which relate to election. But reflection will remove the conception that such words are opposed to the absolute salvation of the world. They are, rather, words which necessarily flow from it. Let the statements which affirm that the world is to be saved through Christ be first received, and all these passages, which seem to us so opposite, conform themselves to it perfectly. There is not even a semblance of opposition in the words themselves. For the eternal is not future: the state in which corrupt men are, is eternal punishment, or ruin, or perdition:* a state from which they are set free by having eternal life given to them. Election is a simple fact before our eyes. Some believe and are saved, others do not believe and are damned: it is God's grace and choice which determines who these are. But this does not prevent the saving of the world. Does not St. Paul, speaking of the Jews, say first, 'The election hath obtained it, and the rest were hardened,' and add immediately afterwards: 'So all Israel shall be saved'? The passages which speak of death as the result of individual sin, such as 'the end of these things is death,' present no embarrassment when the twofold aspect under which men are regarded is borne in mind. The individual's death results from his own sin; the death in which all are is a condition affecting man. Sinning truly leads to death, but it is from death that Christ redeems. He makes the dead alive. The language of Scripture is consistent throughout, and perfectly natural; so natural and true that we cannot help speaking in the very

* Very remarkable is the expression κόλασις αἰώνιος (Matt. xxv. 46), an eternal discipline, or chastisement, a correction in respect to the eternal. κόλασις means pruning, as of a tree. So strongly is the meaning of discipline in it, that it is used for chastened, in the sense of moderate, desires. *Arist. Ethics*, iii. 12.

same words, when we receive it simply as it is. Taking as the basis of the whole that which is laid as the basis, that Christ has come to save the world, and will save it, drawing all men to Himself, the words of Scripture arrange themselves in perfect order and consistency. Death is, and will be, to those who do not believe; eternal punishment will be theirs; they will be cast into a lake of fire: 'the smoke of their torment goeth up for ever and ever, and they have no rest, day nor night, who worship the beast and his image, and whosoever receiveth the mark of his name.'*

What we learn from the Bible is the fearfulness of this state which is ours; this state which men like so well, the evil of which we naturally regard so lightly, and yet feel truly to be so fearful when once our eyes are opened. Shall we never understand that we may be in hell, and like it; that of all evils that is the worst and most to be deplored, and the one of which, if God speaks to us, He must warn us in the terms of deepest awe and most touching love? Speaking of liking evil, must not He say: 'Turn ye, turn ye, why will ye DIE?' He whose Life is shown us in the Cross? How should He speak of *pain* as death, whose heart we see in Jesus? And for the power which should make us flee, the terror by which we should be moved to escape, does not God open our eyes to see things as they are? By showing us Himself, He makes us fear the true evil. He makes us fear that which we did not fear, by making us know that which we did not know. Can he, who believes that he sees God in Christ, and that in HIM is the

* Rev. xiv. 11. These words describe the state of men in this world.

LIFE of man, fail to feel that sin is the most awful punishment; must he not be saved from damnation?

Upon the conception that the world is not absolutely to be saved, no view can be taken that does not bring us apparently into direct opposition with some words of Scripture. So the Calvinist and Arminian systems continue to divide the world, and with no prospect of reconciliation. For each side finds some of its positions clearly in the words of Scripture; each finds other words which it can scarcely embrace. Men are Calvinists or Arminians according to whether they think most of the universal dominion of God, or of the free-will of man; whether they most demand justice or grace in the divine dealings. Their fatal agreement in postponing the eternal to the future, and seeing death in suffering, renders all other agreement impossible. Very striking it is to observe how one man rejects a provision of salvation for all, which takes effect upon some only; not perceiving that his own conception, of a salvation not provided for all, is even more impossible to others, than the thought that any for whom Christ died shall not be saved is to himself. Yet is nothing more simple than the union that seems so far off. Well does the Calvinist affirm that all for whom Christ died must be saved; well the Arminian, that Christ died for the world, for all men. What then prevents that they should unite in affirming that the world must be saved? This only: that both have a certain idea about man and human life; that both think the eternal is an everlasting future, that both think men cannot now be dead, in the worst sense in which the New Testament speaks of death.

The doctrine of election receives beautiful elucidation from those words, 'Strive to enter in at the strait gate'; for those who are saved are not few, but a company 'whom no man can number.' Evidently the words refer to the facts of life: few enter the strait gate, but many go in the broad path which leads to death; for 'she that liveth in pleasure is dead while she liveth.'

The numerous passages, especially in the parables, which speak of the separation of the righteous and the wicked, and the various terms in which the future punishment of some is described, come to our minds as if they were in opposition to a salvation which should be effective in the case of all. Yet why should we conceive them to be in opposition? Why should we array against each other passages which do not clash? If the salvation of the world be absolute, if all men shall be brought to Christ, are those passages the less true? There is no reason to modify their language, or to endeavour to evade their meaning. The redemption of the world from death does not contravene punishment, does not involve a confounding of the evil and the good. Let the absolute salvation be believed; do those other passages lose their meaning, their force, their necessity? Sin shall be punished, the workers of iniquity shall be banished from God; they shall be overwhelmed with 'eternal ruin,' 'drowned in destruction'; yet none the less shall Christ draw all men to Him. None the less is 'the free gift come upon all men unto justification of life.' The absolute salvation of all men is an explicit statement of the New Testament, an emphatic and unequivocal declaration on the part of the very men who affirm the supposed opposite things. Nothing more is needed to prove them not opposed than simply to believe them all.

Nor is it, indeed, difficult to understand why

men, whose aim and work it was to reveal the salvation of the whole race of man, should have dwelt so largely on the doom of the evil, the punishment of sin. It behoved them, above all, not to ignore the divine justice, to treat lightly the demands of law, to seem to confound good and evil. Their part, especially, it must have been, to show that the demands of the conscience were not set aside by the salvation they affirmed, the wrong and guilt of human life not disregarded. How else should they have been believed? There are many who tell us that the world shall come all right at last, that there is nothing so much amiss; but we cannot believe them. Our conscience tells a different tale, demands a different issue. But when the very men who, more than all others, have stirred our conscience, opened our eyes to the majesty of the divine law, the hatefulness of sin, the impossibility of evil escaping vengeance; when they go on to affirm that, with all this, as the sum and conclusion of the whole, the WORLD is to be redeemed from death, and that God shall reconcile ALL THINGS to Himself, how should we disbelieve them? on what ground base our disbelief, what reason allege any more for refusing the Gospel that they preach? They have themselves anticipated all objections, made their own all the ground that might be taken against them. Do we say: that God must punish sin? so have they said: that the holy must be distinguished from the sinful, and a broad contrast of fate allotted to them? so have they said: that God will say, 'Depart from me all workers of inquity'? so have they said: that tribulation and anguish must be the portion of those that do evil? so have they said: that death must be the wages of sin, and the wicked be destroyed with an absolute, an eternal ruin, the opposite of eternal life? so have they said. With all this, ruling, and reconciling, and crowning all, weaving all into one scheme of glory, comes the proclamation of the Christ, the Redeemer of the world; who, first subduing all things to Himself, shall lay down His honours at His Father's feet, and God be all in all; death swallowed up in victory.

Shall we dare to put against this our preconceptions? Shall we disbelieve the Gospel, because we do not see how men are to be saved after the death of the body? What avails our ignorance to overthrow God's word? If we examine ourselves, do we not find that we have moulded our opinion on the whole Gospel, upon the supposition that we know what happens when men die? On this assumption it is that we have converted the proclamation of the redemption of the world, to believe which is life, into an offer of salvation to be accepted before the body's death or lost for ever. But how should the dying of the body prevent men from believing in Christ? How do we know the nature of the change which passes upon men in that so-called death? It is our supposition, indeed, that the Bible affirms that there is no salvation after death, but have we ever looked to see if it be truly so?*

But not only is the salvation of the world affirmed so emphatically, in direct terms, by the New Testament writers, it is the spirit and life of all they say. Allusions to it break out continually, as if it were

* The passages commonly quoted to prove this idea might furnish the best evidence that it is foreign to the Scriptures. 'As the tree falleth so it lieth,' for example. For further remarks on this subject, see Book V. Dialogue 2.

the great subject of their thoughts, the great joy of their hearts, the centre about which their life revolved. It seems the source and consummation of all their message. 'He is the propitiation for our sin,' says St. John; 'and not for ours only, but for those of the whole world.' The former could not suffice without the latter, of which it is, indeed, but an offshoot and consequence. For, that Christ takes away the sin of the world is the essence of the Gospel; the salvation of the individual arises out of, and flows from, the salvation of mankind. If Christ did not give life to the world, it were given to no man. A common death and curse is ours, a common deliverance only avails to save us.

And in St. Paul's language a constant reference to the same thought is seen. Inseparable from the curse and condemnation, from the threatenings and warnings, ever comes the fact of the redemption; because they never forgot the redemption, the apostles never forgot the curse; they were able to assert the righteous judgment of God, the unswerving justice, the awfulness of the damnation, because the salvation was ever present to their thought. Our tongues falter, and our hearts fail, when we would assert the death that is the wages of sin, because we see not the fulness of the life that is the gift of God. The letting go of the absolute redemption benumbs our grasp of all the rest. We dare not speak as they spoke, because we are not thinking as they thought. How could we say with St. Paul, 'Therefore God shall send them a strong delusion that they should believe a lie, that they all might be damned who believe not the truth, but have pleasure in unrighteousness'? Can we picture to ourselves what man it must have been who, thinking as we do of damnation, could have written those words? But if to be damned is to do sinful things, and if from this damnation all men must be saved, why should it not be even so? Why should we shrink from words which do but express the facts of the world's history? God sends men strong delusions, and they believe lies, and are damned. They are damned that they may be saved. Even as God includes all in unbelief, that He may have mercy upon all; or as by the law is the knowledge of sin. For how could man be saved except through sin? Men are not evil because they do wickedly, but they do wickedly because they are evil. Only by the wickedness of our deeds could we learn the evil of our hearts, or know our need of being made new. If the world were not wicked, if men under temptation and delusion did not run into crime, fall into damnation, Christ had died in vain to save it. To have been not sinful, man must have been left in hopeless death.

For the history of the world is the making man alive; that is the resolution of the mystery: Man being raised from a state so evil that he might not continue in it, a state of which God's very being necessitates the destruction. All this sin, all this woe, must be, for death must be destroyed. Man cannot be left the self-regarder, the passional, inert being that he is. At all expense of sinning and of suffering he must be freed. The weary waste of human life becomes quite new when we see what it is for. God's glory is the light which glows even in the lurid flames of hell; one glory with heaven's own brightness, the glory of eternal love. The fire that consumes all evil lights with perpetual day the heavenly city. 'God is glorified alike in those that believe and in those that perish.' He is glorified in sin; even in chief degree glorified in sin; for He bears sin, bears it for man. That evil thing which His soul hateth He endures that man may live. He is glorified in sin, as when a righteous man bears wrong and insult, unavenged, for love. Then and therein is glory, the glory wherewith God was glorified when Christ hung upon the cross, the glory for which He prayed: 'Glorify me NOW, with thine own self, with the glory which I had with Thee before the world was.'

God bears sin for love; for love of man. We see Him in the Man who bore our sins. He shows us therein what He is. No heart is tortured by sin and misery like His, whose prerogative it is to bear alone the name of Love. No bosom throbs so deep with pity, or burns so intense for justice. But He endures for love of man. The infinite patience faints not, nor is weary. Nor hastens; for there is no delay. The slow progress of the painful hours passes not too sadly, the catalogue of wrongs and woes grows not too full. The Father's eye watches the unfolding life, and rejoices over the Son that from the dead is made alive again. For He has shown us how He sees mortal life; the type and pattern of humanity, the Man to whom it was given to have life in Himself, has revealed it. It is interpreted in Christ. Life, His own life, given to the dead; for in Him God reconciles the world unto Himself, not imputing their trespasses unto them.

CHAPTER V.

OF HEAVEN.

Ought not Christ to have suffered these things, and to enter into His glory?

IT cannot be denied that the Bible has been laid as a yoke upon the human heart. If any man think that a repudiation of its authority can proceed only from the evil part of our nature, he has more to learn respecting humanity; he may have much to learn respecting himself. For the reproach has been often uttered, and is widely believed true, that the religion of the Bible is a selfish religion, that its main maxim is to secure our own interests; and men whose hearts rise up against the dogma that self-interest can be the true or rightful spring of human life, condemn it unheard. But is it not taught that the Bible makes self-interest the basis of religion? Is not that reproach inseparable from the doctrine that life and death are happiness and suffering? Let these ideas be refined to the utmost, the pollution of self-love cannot be purged from them. Men will still say, that religion is but another form of self-seeking, and not deliverance from it. The world rejects a Gospel clothed in a garment which makes it but the reflex of themselves. The substitution of future happiness and misery for life and death, of something to be got for ourselves for deliverance from the necessity of self-regard, is the death of Christianity. It cannot rob, indeed, the death of Christ of its saving power over individual men, but it despoils the Gospel of its prerogative, and quenches in darkness the life that should be the light of men.

For the absolute salvation of the world must be denied, if salvation be identified with a future happiness, with the escape from misery or suffering. Necessarily denied; for so conceived the conscience protests against it. That wicked men should pass to a happy future after death could not be admitted; the eternal laws of right demand a different result. The everlasting misery, which seems to be the only alternative, is accepted of necessity. It is our nature to think of the future in this way, to conceive it as the scene of enjoyment or of suffering, and to erect our system of religion on that basis. That is a natural religion, based on our conviction that the eternal life must be, in its essential elements, such as ours is now. We stipulate for ourselves that we shall be happy, that we shall have enjoyment. We hold forth to men that heaven consists in highest, purest pleasures, but in pleasures still.

A very different thought, indeed, is in the heart of many who hold this language. By the happiness of heaven they mean love only; but the doctrine is that which must be spoken of: it is the doctrine, not the feeling, that is presented to the world. Men do not hear of heaven as love or life, but as happiness; of hell as not loving or death, but as misery. This is the Gospel that man rejects, and does right to reject. The world thirsts and groans for life; it is weary and in despair for lack of love; but a future happiness it will not have, a future misery it does not fear. With desperate and wild resolve it clutches the present; with a madness no experience can tame, men pursue wealth, pleasure, glory; the undying worm within them cries for ever: Give. That which is now is that which is for them. 'Tis right: the voice of God within them bids them cleave to that which IS. The present is the eternal. A false, deceitful voice it has been, that bade them postpone that which is now to that which shall be; a wise defiance they have hurled against it. If the world would have accepted a salvation from that which is future, it could not have been saved; for its ruin is eternal. The grasp upon the present is a mute protest for the eternal against the temporal. But to hold to the present is not to take the seeming for the fact. Really to cleave to that which is, men must know that which is; they must know the eternal. Men are deceived; they perish for lack of knowledge. God has sent them a strong delusion, that they should believe a lie, for they believe that the eternal is not now; that things which perish in the using are the facts that ARE. In very truth they do believe that a man's life consists in the abundance of the things that he possesses; that life is not eternal life, but transient pleasures, pursuits, activities. They do not KNOW, and thus are lost and ruined, for phantoms sway them as realities; they make their LIFE in that which does but seem. Therefore to save them, God shows them that which IS, the true eternal fact which they see not, although 'tis all around them. Himself He shows them, His own being, His own life, partaker of their death, that they may know the eternal in utterest sacrifice of self.

Knowing what salvation is, it no more does violence to the conscience that all men should be saved. It subverts no justice, that the power of

Christ's love should subdue in every man all self-regard, all the force of sin and of desire: that, in the redemption of man from death, all men should be made complete in love, and, striving with death, should utterly sacrifice themselves, and endure like Him. For this is the salvation which God bestows: to be one with Christ, who while He trod the earth was in the bosom of the Father; and shows us heaven in sorrow.

Christ saves us, not from suffering, but from death; not from pain, but from that which makes us flee from pain. The men whom Christ has saved are known for eminence in suffering. The stream of life runs red with blood. 'They were destitute, afflicted, tormented; they wandered in deserts, and mountains, and dens and caves of the earth.' To them 'it was given to suffer for His name.' Life is, not to be afraid of suffering: to be so perfect in love that loss and sacrifice, extremest and utter sacrifice, the having nothing, but giving and bearing all things, is perfect joy. A life which we cannot fully have while we are in this earthly state, while man is wanting in his life; but by the bestowment of which on man his deadness is taken away. For herein differs the physical from the spiritual, man from God; that to man sacrifice is pain; love brings suffering, that which we dislike and would fain avoid. We groan, being burdened with this weight of flesh, which, making us feel the work of love as evil, puts perfect life beyond our reach. For to God, and to us when man is wholly made alive, that which we call sacrifice, the entire self-abnegation, has no pain. There is nothing to which the loving act is unwelcome. For the character of our present state is, that the action which love prompts is painful; there is that in us which is opposed to love. Here is the evil of our condition; we like that which is not good. When man is made perfect, the defect removed from him, then shall that which is now painful to him be no more painful. Perfect sacrifice is heaven to those in whom love is perfect. Not in our circumstances can be the change from earth to heaven, it must be in us; in the taking away that deadness by which man is as he is. Then is perfect sacrifice, solely and only giving, no more a pain. We cannot think it, but we know it. In our hearts we know it. Not in vain has Christ shown us God. Our thoughts are truer than our words; that which we believe and hope for, than that which we profess. In heaven we look not for enjoyment but for love. Only certain intellectual conceptions, notions which we have formed on abstract questions, interfere. Our speculative ideas make us speak of heaven as we do not feel, and the world laughs us to scorn. For the Gospel of happiness and misery is not true to the heart of man. It does not touch the strongest chords in human nature, the true movers of human life. Men know indeed that they love self, that they are guided by self-regard and pleasure, that they do seek what they like; but they also despise themselves for it. They also utterly contemn that life, and treat it with bitter scorn. That is not humanity. It should be no strange doctrine to man to hear that he is dead; who could have said it more plainly than himself, or in words of deeper mortification and despair, scarcely veiled by the thin disguise of mockery?

It is not true to the human heart, it is not fair to man, to come to him with a religion that concerns his happiness, his escape from suffering. Such a religion cannot save the world. Can he be saved from himself? Can he be made different in his inmost being, raised from regarding his own pleasure, from seeking his own interests? This is the question religion has to solve for him. The question for humanity, this day, concerns the resurrection of the dead.

Is not that which God gives to man in saving him, in making him alive, the power of giving; of true and absolute self-giving like His own? Is not that our *want*, truly what we long for, and yet do not know? That is the Eternal, that is God's own life. That is the water, drinking of which man shall thirst no more. For it were mere stagnation and satiety, to get so that we should never want again: that were the end of all enjoyment, it were merely mean and small capacity. We are to be delivered from the necessity of getting; from that which makes hell, and is the very bond of death. Getting and enjoying have been tried all these ages: amid God's richest gifts, amid His infinite bounty, man has lived to get, and the result we know. In no form or way could that continue everlastingly. Can we be made givers, be made alive? To be made different in ourselves, so that we shall not be for ever grasping for our own, that is the true, the eternal life.

We suffer ourselves to speak as if the cause of our evils were in God, and not in ourselves: as if our unhappiness were from the poorness of His gifts, because He has surrounded us with conditions so imperfect, with pleasures so unsatisfying. But it is not so. God's benefits know no abatement nor increase. He gives us all things richly to enjoy, crowning our life with loving-kindness. Man banishes the glory; and seeing not, thinks darkness is around him. Heaven has no other joy or glory than is now, but man shall be different: active where now he is passive; partaker of the life which now he feels to crush and to subdue him. And feeling all things differently, no more compelled to want, the unutterable bliss of perfect love is his. For heaven nothing must or could be changed but man. Nothing but the death destroyed or taken away. Not less of sacrifice, of being utterly given up: not less of love without us, only more love within.

We deceive ourselves if we think that altering the form of our getting could make a heaven; the self would be our torment still. 'Tis not the things we have to bear that crush and ruin us; it is our necessity to get, our want of something for ourselves, our constant craving. That is our perdition. God must give Himself to us. He must be in us. His life be ours. So we shall want no more; have no more emptiness to fill. We shall be like Him, able to be content with giving. There shall be no more want. The infinite life shall fill us; the absolute love and sacrifice, in which alone eternal being is, shall be ours, shall be enough for us. Evil can be to us no more, nor sorrow; for all sacrifice, all giving up, all that is now enduring, resigning, bearing, being tortured, set aside, and downcast, then shall be our joy. The death that makes it pain shall have been done away. What is it that never faileth, which alone vanisheth not away, but the love that endureth all things, beareth all things, seeketh not her own? This shall never cease. Would we put away self-sacrifice from heaven? It cannot be. Heaven must be self-sacrifice made perfect; ceasing to be sacrifice, only because complete.

Therefore must Christ have been a man of sorrows; therefore must He have borne our sins and carried our sorrows, and taken on Himself the chastisement of our peace. He had to show us God, to make us see what He is, to make known life to death by life in death. In that which we call sorrow and humiliation and sore distress, all things borne and sacrificed for all, God is seen, and only can be seen by us, while we are as we are, and love brings pain with it. Heaven is, to be and do what Christ was and did, and find no pain in it, because no more is there in us anything that is opposed to love.

CHAPTER VI.

OF THE RELIGION OF NATURE.

The kingliest King was crowned with thorns.

HEAVEN is happiness, the deliverance of man from all that makes him subject to sorrow; but not, therefore, can it be sought as happiness, for happiness is in the leaving off that search. There is no happiness like that of love, but it cannot be obtained by seeking. Love must be given us, must carry us away, must become our nature and our life. That which makes the happiness is gone when happiness is sought. Is the love of God less sacred than that of wife or friend? Can He accept a love which friend or wife would repudiate with disdain? We are compelled to love Him; that is the overmastering passion of our souls, the joy of our hearts, our life, and breath. Such passion we have felt faint traces of before, when most of loveliness has moved our souls to ecstasy, and bowed our hearts to worship. When we could not contain ourselves for joy, and earth was glorified by one presence everywhere, then was revealed to us the image and the shadow of the love of God. We love Him; it is the one fact of our life which flows into all things. To love Him is but to know Him, but to awake from sleep, to have sight given to our eyes, life to our hearts. One presence glorifies all the world, for all things are the presence of Him we love. All passions in this passion receive their fulfilment, reveal their true meaning, become absorbed, and die into life. All human passions mean the love of God, but men know it not. They clasp in ignorance that which fades and passes and is not, not knowing the eternal joy for which their souls cry out.

How should love be spoken or explained? We love and are happy. We do not want, we do not pursue. We want to be, to do, that which He wills and does. We want to give, to bear, to sacrifice ourselves. We love the Infinite, the Eternal, Him in whom, and for whom, and to whom are all things, whose will is done in heaven and earth. His will is our will; we have nothing to get; we love Him. For He is the Redeemer: He takes away our iniquities, delivers us from sin, and will make us perfect and unblameable. He takes away the sins of the whole world, and will be death's destroyer. The Vanquisher of evil by good, who shows us what it is to LIVE; how should we not love Him? How should we cease to rejoice and to be glad in Him? The joy flows too full, it overflows and carries us away.

It is difficult to love God because we do not see that He saves the WORLD; and this we do not see because we make the New Testament express our own natural conceptions, instead of suffering it to speak its own language. It is saying truly that this is man's death, but we are thinking falsely that it is his life; and interpreting its words according to our conceptions, we find in them a confirmation of man's thought of the world, instead of a deliverance from it. We make it even below ourselves, and cannot escape from the feeling that we could conceive a better history for man than

God in his sovereignty has ordained. It is this crushes our affections, paralyses our hearts, makes our piety so lifeless. God's redemption is the making MAN alive; the making all men perfect in self-sacrifice, uniting them to Himself: not some men saved from future misery, but all delivered from this eternal death. Of the true death, the true damnation, He is speaking, and of the true life, while we are thinking of that which seems to us, of that which we like or dread. We think Christ came to give us that which we most wish for, happiness; to save us from suffering, which we most fear; but he came to give us a better gift, to save us from a worse evil. He came to give us life, that we might feel the true good; light, that we may see the true evil; to cure us, that our desires may be different. Christ is the physician: He heals the sick humanity, allays the burning fever in its veins, calms its delirious passion, dispels its dreams, soothes its mad appetites, satisfies its wants, restores it whole and in its right mind to God. He gives us LIFE. With Him we are raised up from the dead.

Our natural conceptions are of no weight, because man is wanting. We are no standard; it is we that have to be altered, a different existence is to be made ours. Never should we use the words which say that man was made in God's image, without remembering that the same words affirm that man is not as God made him. We look in vain for God's image in ourselves. That is to seek the living among the dead. Not in our powers, our arbitrary will, our reach of thought, our dignity of virtue, not in any of these things is God's likeness to be found. Show us thy glory, O God, that we may see Thee. Come forth from the clouds and darkness that are about thy throne, that the resplendent light may scatter all the shade. We will gird up our hearts to bear the terror, our shuddering souls shall face the awful splendour. Let the majesty of thy power be seen, though it crush us and we die.

God doth reveal Himself:—a Man hanged on a gallows. It is too much, O God! Art thou so much above us; are we so unlike thee? is this the power whereby thou art able to subdue all things to thyself; all power in heaven and earth given unto love? O foolish heart to tremble and be afraid of God! O idle sense to heap up images of vastness! O proud and evil thought that God was like us, only more. Well might vague terrors haunt our souls, and secret dread, and enmity, and wish to hide, for we thought God was like ourselves. Unlike us most of all in this, that while we turn away from our own image, erected to us for the God we do not know, HE seeks us, loves us, will not let us go; will make us know Him, will make us friends. For to know HIM is to love.

We cannot see God as He is, for we see ourselves instead. We cannot see Him in nature, for we put our own deadness into it. We draw our thought of God not from that which IS, but from that which we feel to be, and make Him a self-seeker like ourselves. God is not to be seen in nature, as we see it. The fact would teach us God, but the phenomenon will not. He is such as the true being of nature would show Him; not such as we infer from nature as we feel it. We do not see love in nature; we persuade ourselves we do, and so make our dim vision dimmer still. The love we fancy there is such love as ours. Not even such. Nothing can more pervert the religious sense than calling what we see in nature the love of God; this ruling providence, this exercising of power, this giving enjoyment, even if it were much more perfect than it is. This is man's love; this is our way of loving, as far as the earth is beneath the heavens from any that can be called divine. There is no sacrifice in it. 'Tis like a rich man who, happy and comfortable himself, takes pleasure in making his dependents so; keeps good order, protects virtue, and does, without trouble, everything benevolent. It corrupts the heart to think of God's love so. If we would truly see God's love, we must seek it where Christ sought it, in sorrow, and sin, and agony; in that which is wretchedest and vilest. Not in the beauty and delight of earth, which cost Him nothing, but in darkest woe and fearfulest despair. God's love is seen in sacrifice. When we can look on nature thus, seeing in all that is saddest and most evil the fact of perfect and intensest love; life given for the dead; when we can interpret nature by the cross of Christ, then we see God in it; but not till then.

We cannot express it, God forbid! That were not love enough; how could that be infinite which man could say? and what need to say it, what scope for words, what place for idle breath to draw faint images? The fact unutterable, inconceivable, is here. In death endured for man, God's own heart wrung with human agony and bowed to willing shame, there and there only can we see the love of God. So HE loves. Let us not profane that holy word, the love of God, which, seen even so, is but darkly seen, as man's fainting eye can bear; not the true brightness of that glory, but the image veiled and softened for our sight; let us not profane that word to aught less worthy. Till we can see nature as one with this, we do not see it; we search in vain in it for God. 'Behold I go forward, but He is not there; and backward, but I cannot perceive Him; on the left hand where He doth work, but I cannot behold Him; He hideth Himself on the right hand, that I cannot see.'

The secret of the universe is learnt on Calvary. In that death the life of nature is revealed. For it is love, perfect and utter sacrifice; it is the life that shall be ours when man is made alive. Then deadness shall no more be felt around; love shall know love, and life within shall answer to the life without.

CHAPTER VII.

OF FREEWILL.

Love is the visible form of freedom.

THERE is nothing of which men are more conscious than of their failure to solve questions which are of the greatest interest to them, a solution of which would be received as an invaluable boon, not only for theory but for practice. There is no necessity of their condition to which they submit more reluctantly, than to this enforced acceptance of that which they cannot understand. It is not too much to say that no thinking man is satisfied: the cherished expectation of knowing more fully in a future world sufficiently proves the feeling of loss and want, which the acquiescence in mysteries does but disguise.

It were an infinite joy if the mystery could be taken away from our life, and we could know the true answers to those earnest questionings, which men from earliest to latest times have addressed to themselves and to all things around them. No heart would not beat high at the prospect, would not purchase it at any sacrifice. The promise rouses now indeed but a feeble interest; for faith has grown weary, and disappointment by long use passes for success. We gild the chains upon our hands; almost we have persuaded ourselves that they bespeak no slavery. But the true heart is not deceived. The fettered limbs are motionless, that the galling be not felt: the frantic struggles to break free have given place to calm; but it is the calmness of despair, that mocks itself with laughter, and hides its writhing agonies with boastful words. It is despair; that mimics trust, that compels itself to resignation, that smiles the bitter smile of scorn, that despises, disbelieves, or believes because it will believe, while the cruel doubt within, the cruel sin without, torture the faith they cannot slay. Despair—intensest, bitterest despair—drives its pale victims in the path of pleasure, pursues their steps with madness of desire, gives them no rest day or night, seeking happiness for ever. Despair proclaims its gospel: take care of yourself, obtain happiness, flee from pain, ask no questions about the world, but secure your own well-being.

How can this be the life of man? how can his life be in getting? when the true life, that which he wants, is the opposite of getting, is in self-giving only. How can the remedy for human ills be in the persuasion that we cannot know? Life is to know; death broods over us in this unnatural calm, this falsely called content. Doubt about God, doubt for man, doubt of absolute right, of the perfectness of love; the dread that there is something which is opposed to these, which no light could make clear, no better knowledge reconcile; these are the things which eat into our hearts, and leave us no alternative but reckless strife or deadly apathy. Not doubts that we would express in words, doubts that we crush like sins, and deny, and prove that they ought not to be; but doubts that are the worse for being crushed, the blacker that we will not face them, which poison the very springs of piety and make it false even of love to say, 'She seeketh not her own.'

All comes from measuring God's work by ourselves, taking our feeling as the standard of that which IS. There is no more doubt if we will verily believe that we are deceived. The secret of man's perplexity is, that he believes himself. Feeling himself free, he affirms his freedom, asserts his life; but Christ's words affirm that he is free only when God makes him so, dwelling in him and giving him a life he had not. We have mistaken freewill for freedom. The deception we have been under, its source and necessity, the whole history of thought on the subject of freewill, are transparent, when we recognise the central fact of human history, that man wants life. True freedom belongs to manhood; the freewill of which we are conscious belongs to death. Freewill is not denied in denying man's freedom; but freedom is asserted to be a different thing. God is free, to whom sin is impossible. Man is free when sin is impossible to him.

For if our conception of freewill be analysed, it will be found in itself to indicate, and correspond to, a state of defect; the essence of it is, that the action should be not necessary. It is not necessary to man to do right. It is worthy of remark that the ideas of rightness, or holiness, and of wrong, are differently related to necessity. For while necessity excludes sin; and an action which is necessitated loses the character on account of which we can attribute criminality to it, the case is not the same with holiness. We cannot think so without blasphemy. The highest holiness is necessary holiness. Necessity is wanting in respect to man. He is not therefore free, but he is conscious of freewill. He is under law, and justly amenable to reward and punishment. When he is freed from this state of defect, necessity will no more be wanting to his action. He will be holy even as God is holy; no longer liable to sin; free, and controllable by punishment or reward no more.

Reward and punishment may be regarded as serving to supply the defect of necessity in the being to which they are applied. They have evidently this tendency, though but imperfectly. They tend to insure that rightness of action which otherwise might be wanting. Technically speaking, freewill, and all that belongs to it, is a result of 'negation.' Man differs from God in being not free; or by absence of that which in the Divine nature makes holiness necessary. Whatever it be, that is not in man; and man's characteristic of possible doing either right or wrong evidently depends upon that absence, is wholly accounted for by it, and needs no other supposition. Surely God is truly the Free. He must be taken as the type of freedom. In what respects we differ from Him, so far do we depart from the true perfectness of freedom. And in this especially we are unlike God, that it is possible to us to do wrong.

What makes a difficulty here is, that we do not feel as if our freewill, our arbitrary power of doing or not doing, depended on a defect. To us it seems a great capacity, the prerogative by possession of which we are exalted above all the world. But should it be so hard to us to understand that we put darkness for light, and light for darkness? We pride ourselves on arbitrariness, which God puts far away from Him, which freedom knows but as her direst foe. Our own words reprove us.

We labour under a confusion which has its source

in the belief that the world is at it appears. We feel that we are free as compared with the inert things around us. This is true. These things are subject to an inert necessity; man feels himself not to be so. And he dares not allow himself to be not free, for fear of reducing himself to the level of those things. But this perplexity is removed when it is remembered that these things are but phenomena: that there is no inert necessity, nor can be, that it can only appear. In maintaining his freedom man denies his subjection to the phenomenal or inert necessity, to which he is conscious that he is not subject; but he overlooks his want of the true not-inert necessity. The inertness in that which man perceives, and the arbitrariness in himself, are correlative; two forms of the same negation or defect. Nature is free, man is not free, therefore he feels himself arbitrary, and nature bound in fate. Nature's holiness to him is inertness.

With respect to the attempts which have been made to bring man's actions into one category with physical phenomena, and prove him subject to an inert necessity (whether metaphysical, by logical arguments, or inductive, by accumulation of statistics), one remark may suffice. Whatever the force or value of these proofs may be, they are evidently incomplete. The theory does not fulfil the conditions of a theory, for it does not account for the phenomena. It is true that consciousness is not authoritative, but it demands to be accounted for. This is a claim that cannot be foregone. Any theory of man's action, that can pretend to correctness, must show why our consciousness should be such as it is. If man be the subject of an inert necessity, as phenomena are, why does he feel as if he were not?

It accounts for all, and answers all demands, to recognise man's deadness and his presence in a living world. Life and freedom are one. What man recognises in his feeling of freewill is a difference between himself and nature. He is not deceived in this; he errs only in interpreting his consciousness into proof of his freedom. By his deadness it is that nature is to him inert; that the necessity in it appears to exclude action. For truly action and necessity are one. What man calls his freedom, that very thing is his bondage. His self-action is inaction. For what is sin but the absence of the true action of the man, when he is swayed by passion, led captive passively? In sin there is action which is perverted and evil, because the MAN's action is wanting in it. All nature's part was perfect, but there was a demand for human action, and it was wanting.* Is it not wonderful that piety compels us to take all evil action to ourselves, and to say of all good that it comes from God, that it is God's work in us and not our own? Those words are true. Only when God acts in him is man truly free.

That is true freedom in which the action is the Being's own, wholly determined from within. Such action is one and invariable, is necessary; the Being does not change, nor, therefore, the action. True freedom involves holiness. Action that is variable, that may be either right or wrong, is dependent upon circumstances; it is the inaction, not the true action of the Being. This arbitrary action is emphatically not freedom. True action, true freedom, true necessity, true holiness, all are one.

---

* 'I dare do all that may become a man;
Who dares do more is none.'

Necessity and freedom are one in love. True life, or being, and love cannot be separated. It is not a mere figure of speech that God is Love.

Thus the idea of responsibility takes its right place. Doubtless man is responsible; in that way his want of life expresses itself. By absence of the life which constitutes the true necessity, arises that want of necessity which places under law, and gives rise to duty. Law is from the absence of love. When that which fulfils it is not present, then law is felt. There is then, and then only, that which is 'due,' because it is not rendered; that which is 'owed,' because not freely paid. The law is holy, just, and good, for it expresses love; but love in relation to the not-loving. It is not a condition of the perfect life; it is the form of love where the fact is not. Not the relation in which the living stand to the living God, but a relation from which God redeems by giving life to the dead. Law is latent, as it were, in love, like an inscription in a fountain, to be read only when the stream is dry. Law fulfilled is destroyed and done away. Love blots it out: for it cannot save; it brings sin to knowledge, and leads to the Redeemer. Law reaches not eternity; it has necessary reference to time, and presupposes fear. Subjectness to law marks the difference between ourselves and nature. We dream of laws in nature; not law is there, but liberty or law fulfilled. We are under law who boast of freedom, and are slaves. For man and nature differ, as does a dishonest from an honest man. Nature will certainly do the thing that ought to be; under whatever variety of form, through whatever changes, infallibly that one thing shall be: the form may change without limit, the fact never. But man may, or may not, do that which ought to be; he will do it if he likes. Nature wants no laws; the love that is her life necessitates her being, she cannot be other than she is.

A fond delusion it is that finds freedom in our ability to sin. Yet a necessary one. Man's freedom cannot be given up but with his life. Freedom belongs to man's life, and must be maintained by those who do not see his deadness. But the denial of man's freedom imperils nothing if that be recognised. To disprove man's freedom is to give the crowning proof that he has not life. This is the work in which those have been engaged who have assailed the doctrine of freewill. They have been proving on behalf of Christianity the death of man; laying anew the foundation on which the Gospel is erected, which Christian men have razed with pious zeal.

Therefore have there been the strong arguments against man's freedom, which we have so resented and deplored. Therefore have we been compelled to say that freewill was a mystery not to be submitted to examination. These things were to teach us to be wiser; to wean us from that strong delusion that man has life. For the argument against freewill, that seeks to establish in man an inert necessity, answers precisely to idealism; has the same strength of logic, the same weakness of conclusion, the same inevitable rejection by mankind. Each brings us to a result which does not account for the consciousness it undertakes to explain; but each also proves an error in our thought, and contributes its share to demonstrate that false feeling in man which makes evident his defect. As this life of man, this life to that which is not the true reality, is not truly life; so is the freewill which is connected with it, the freedom relatively to these things, not true freedom. Man is relatively free, therefore responsible; absolutely not free, not free in respect to the eternal and truly real; therefore to be redeemed and introduced into liberty, made partaker of God's freedom as of His life.

CHAPTER VIII.

OF THE SELF.

We are such stuff as dreams are made of.

IF our self-action be not true action, what is our Self? of what are we conscious? We are conscious of defect; man's consciousness of self is the feeling of his want of being.

There is in this conception nothing abstruse, or remote from our ordinary modes of thought. It is as familiar to us to be conscious of defect, as of the contrary. When we are conscious of cold, for example, we are conscious of defect of heat. There is a certain heat natural to the human body, defect of which we are conscious of as cold; or in many other respects we may have consciousness of defect; of weakness for example, or of the loss of a member, or a sense. So we might equally be conscious of defect of being—of want of the true being of man within us.

Man is defective. This we know. The perfection of man has never been maintained by any. Therefore, being conscious, man should be conscious of being defective; conscious, therefore, of defective being, and of defect of being. To be true, man's consciousness must contain this element of defect. And can it be doubtful that it is this element of his consciousness to which the name of Self has been assigned? Is it not an emptiness, that we are conscious of within, and call it Self?* If it be asked, what then is conscious of self, if the self be defect? it may be answered: man is conscious of it. Man is conscious of defect of being. We constantly distinguish, in our language, between the man and the self. We say: I hate myself, despise myself. We are deeply conscious that the self is not the true manhood, though obscurely, not having distinctly made the question a subject of thought.

'There *is* the Self.' True: even as there *is* a shadow. Why should we not as well be conscious of defect of being, as perceive defect of light? Only by experience and long inquiry does man arrive at the knowledge that a shadow is a negation, and understand that the effects produced by shadows are due only to want of light. He perceives shadows as things. Always the absence of anything is felt as an existence by us, while we do not recognise that of which it is the absence. Nay more, even to our instructed eyes a shadow will often appear to be the substantial portion of an object; and the important part which a mere negation may play in our experience, may be understood the better by a reference to the painter's art; for drawing is little more than a correct use of shadows.

We are conscious of our self as acting; we recognise actions of the self. Truly: but we also recognise the action of cold; yet we admit that cold is but absence of heat, that the 'actions of cold' are the effects of a negation. Does not our

* The unconscious suggestions of language are very striking. We speak of 'self-consciousness,' familiarly, as a defect in a person's character.

experience teach us that the actions of the self are the effects of want of action in man? Is it not thus that self-will is slavery?

It should be remembered, however, that to affirm the self, of which man is now conscious, to be defect, is not to repudiate individuality. Individuality does not depend upon the self. There is in us an emptiness where there should be a fulness, individual defect instead of individual being. May we not look forward to the true individuality of man in the bestowment on him of a truer life?*

Recognising in the self, of which we are conscious, defect of being, a light diffuses itself over all our experience; the whole thought of man and of the world becomes transparent. There is a defect in us which we apply, as it were, to all things. In all things, as they are to us, our self has left its mark, defect is introduced, the being is banished. Remembering this, it is no longer a mystery that we have been so perplexed and baffled. So must we have arrived at a knowledge of the truth. To say that the things that are real to us are not the true realities, is but another way of saying that our self is not truly being. Thus too we see why, in our inquiry respecting nature, we must inquire also into ourselves. How should that which a defective being feels, fail to indicate his defect? Must it not differ by defect from that which IS?

If this self be defect of being, then must all true good, all life to man, be in self-sacrifice. In the utter destruction and casting out of this self, the

* We may either say that the self, as such, is defect of being, or that the self we are conscious of is not the true, right self, which is 'being.' Whichever be preferred, the idea is the same: the argument refers to this self which is in us.

doing away of the defect, man's life is given him; there can be no other true life for man. And when man truly sacrifices self, it is God's act in him. From the self comes no goodness, it must be made new. Self-righteousness is not righteousness; only when our self is filled, our emptiness destroyed, by God in us, then does man live and act. That is the true humanity. That also is man's true freedom. Sin, as the assertion of the self, is the bondage, the death of man.

Thinking of our self as 'being,' taking the self-view, we are lost in darkness; but the right knowledge of our self enables us to reconcile almost all contradictions. We enter into the mystery of man's condition: that he is so glorious, yet so mean, so elevated, yet so ignoble, has such capacities, yet effects such unworthy results, is the one thing which seems to come short of its destiny. The true being of man is not in this that we call his being. Humanity looks towards, and demands a different being from this. This self is not man; it must be destroyed and taken away before the true man can be. There is demanded, in respect to man, an EXISTENCE that is not yet; his nature, his feelings, his consciousness are attuned to a different being. He is conscious of defect; conscious of it, because it is to be supplied. For thus the statements of the New Testament are felt to be self-evidently true, in proclaiming MAN's salvation; that this defect and death of man shall be done away; God shall be all in all. It must be so. How should man be conscious of defect, except that he is to be delivered from it? What is it to be conscious of defect, but to have an aspect towards a truer being, a relation to a state of freedom from defect? So the true relations of sin are seen; it arises from the self, and exists for its destruction. For human experience is the destruction of the self, the doing away of the defect in man, the making him alive. Because it is being destroyed, the self runs into such excess, such madness. Driven by fiends, burnt up by torturing fire, it writhes and struggles in its death agony, grasping at every pleasure for short-lived relief, rushing into insanest riotings, consuming itself with known and deliberate pangs, because ungovernable cravings gnaw its heart. Around it sweep the everlasting flames, the wrath of God filled up, and sparing not.* For in sober and literal truth this is hell. The self makes hell for itself, nor can escape, nor shall. 'I will be thy plague, O! death; O! destruction, I will be thy destroyer.' Sinning is hell, the burning up of the self with unquenchable fire. That cruel mystery of sin is this: should not death be destroyed, should not man be made alive? Let the fierce flames of passion have their course, let the weeping and wailing and gnashing of teeth go on, let all miseries be incurred, let all vain remedies be tried, so shall God's remedy at last be found the only one, so shall Christ draw all men unto him.

For by life only can self be cast out, only by self-sacrifice, only through the love that is in knowing God. Thinking of that which we feel to be, as if it were that by which we must judge existence, we do not see that a change in man's BEING must be a spiritual change. The want of life whereby he is in a world not spiritual must be done away. To be in the spiritual world is to be spiritual, to have life, to be free from that self with which we now are

* 'And I saw another sign in heaven, seven angels having the seven last plagues; for in them is filled up the wrath of God.'—Rev. xv. 1.

burdened, 'the body of this death.' For well has it been observed, that our consciousness of self is that which determines our state of being, and gives the character to our perception. This self that we are conscious of makes the world inert to us. Our present self-consciousness demands as its correlative, and condition, an inert existence around us, which passively obeys our exertion, and is respondent to our force. Self-consciousness, involving the sense of exertion, is inseparable from a feeling of passiveness in that on which we act. In our conceptions of the future state, the same element is necessarily present. We cannot conceive of our*selves* except as in a world thus passive. However we may hold that world to be spiritual, though we may even say it is apart from matter, this characteristic of inertness cannot be separated from our thought. We give to that which is thus inert, indeed, emphatically the name of substance; and are but too apt to fancy active powers inhering in such a substance, even when we think of spiritual being. This self of ours carries inertness inseparably with it; where it is, inaction must be perceived. No stronger proof could be that it is the negation: put it wheresoever, surround it howsoever, a negation must be present to it. At least we may say this: inasmuch as the self that we are conscious of, by its own nature, involves the feeling of inertness, therefore our feeling the world to be inert can be no evidence that it is truly so.

From this necessity under which we lie of feeling inertness around us, evidently has arisen our conception of the universe as an inert existence on the one hand, peopled by active beings on the other; and these beings, also, we conceive to be conscious of defect, to be such as we are.

Even God we have conceived as such a self. Scarcely can we prevent ourselves from attributing to Him intellect such as ours, exertion of force like that we feel, difficulties, contrivances, ideas. All this has been necessary from our taking our Self as the standard of being; not reflecting that our mode of consciousness involves a consciousness of defect, and that, therefore, there must be a consciousness different from ours. This self constitutes us physical: to be not physical is to be free from self, to be free from defect. In being made spiritual man is made to BE; self-action, or arbitrariness, is done away; necessity is put within.

Thus the question of man's relation to God, how he can be distinct from the Divine Being, becomes free from the difficulties with which it has been felt to be encompassed. It no longer even appears to be a contradiction to say that God is infinite, and yet deny that man is divine. Man is not divine, because his self is defect of Being. God is not in this humanity, not because He is limited, but emphatically because He is without limits. Of Him no darkness, no negation can be affirmed. The defect in man, our consciousness of a self which cannot be divine, has been truly the cause of our entertaining side by side in our minds the beliefs that God is Infinite, absolutely without limits, and yet that there is other EXISTENCE besides Him. The direct contradiction between these propositions is evident; and though it does not therefore follow that either ought to be abandoned, yet it is at least clear that there must be a juster and more adequate way of regarding the subject, if it could be attained. If contradictions are sometimes necessary to us through our imperfection, they certainly cannot be in the things themselves. Does not the recognition of our self, as defect of being, do away with the necessity of maintaining a contradiction in this case? Our self is emphatically not God, for it is not being. We had to reconcile the feeling that our self was not divine with the false opinion that our self was being, and this confused us; made us assert a BEING not divine, while yet we could not renounce God's infinitude. How much, not only philosophy, but religion has suffered from this paradox, those well know who have studied the history of human thought. But in recognising what man's self is, the disturbing force is taken away, and our thoughts right themselves. No contradiction is here any more, or is to be feared. For in asserting that man's self is not God, we do not contravene His infiniteness, but assert it. And the religious feelings are relieved almost as much as the intellectual sense, while the language of Scripture receives the most striking illustration. To be not divine is man's death: what he wants is to have Being in him, to be united to God, and apart from Him no more. And the Name, The I AM, The JEHOVAH, becomes full of a new meaning, a new glory. God is THE BEING. And not less do the words of the New Testament reveal their true force. Is not this its doctrine throughout, that man's life is, to partake God's life; God to be in him? We need only to give up the persuasion that our self is Being, to see a new and awful meaning in the familiar words. Have we not warped the New Testament to our thought of man's life apart from God, while it affirms his death?

Man's death:—his self defect of being. Surely these are the same. In consciousness of this self surely man is made conscious of his death; conscious of death, because he is to be made alive.

And our thought of God also loses a great part of its difficulty. Ever the battle is renewed, on one hand or the other: Is God a Person? If not, He is nothing to us. We must have a Person for our God, or we are without hope in the world. But the difficulty in maintaining this lies in our taking our self as the standard of personality. God is not such as man: surely not; no such Self is in Him. Falsely we call ourselves PERSONS. We want personality. Then first are we truly personal when God fills us with Himself. And God is not A Person; one among many. God forbid: He is THE PERSON. Then are we personal when we are divine: when the over-mastering Spirit dwells within us and ACTS, and we can say, 'I labour, yet not I.'

What dream is it from which we shrink, of being absorbed in God? as if to be one with God were loss instead of gain; as if our self were Being that we should fear to lose. To be divine is to be personal, to be in the true sense man. Least of all should a Christian man have feared to be made one with God, for what is shown us in Christ, but the perfection of humanity in oneness with God? If Christ be divine and yet human, why may not we be human and yet divine? The notion of 'absorption' bears self upon its face: we think of God as physical.

And if we say, how then can God *create*, if He be the only Being? would it not become us rather to keep reverent silence, than to suppose that creation must conform to our conceptions? Should we not rather learn what creation is from facts, than insist upon a creation answering our ideas?

Why should not creation depend upon the true infinitude and soleness of God's being? Why should we allow ourselves for a moment to think the contrary? If, moreover, we admit creation inconceivable, can there be a greater folly than to assert what its mode must be? And yet again: If man have his true life only when God dwells and acts in him, may we not be well content to believe the same of all His creatures?

The applying physical conceptions to the Divine Being is the secret of the difficulty that has been felt here. For in truth a just thought of the Creative Act seems not so impossible when we remember that God is Love. To our thought love must be self-sacrifice, because of the defect that we are conscious of. Love must be the sacrifice of that which is in us: where death is, life must be its destruction. In self-sacrifice, therefore, we must find the truest conception of creation. Love, sacrificing self: God limiting Himself as it were, giving up Himself for the creature's life; in this most truly may we present to ourselves creation. As Creator, not less than as Redeemer, is God revealed to us in Christ.*

In denying our self to be Being, the relations of things are left untouched. These things that 'are to us,' still are to us. Here is this life of ours,

* Theoretically, it seems simple to say that creation is by negation, not by addition. From infinite Being, by infinite variety of negation, infinite variety of being; that is of relative, or creature being. But such theoretical statements are of little value, except for the purpose of excluding worse ones. They should never be demanded, nor valued, even as approximations. The intellect can only deal with human conceptions, not with the very fact of being.

such as it is, it is not denied. When a shadow is pointed out to be a shadow, an absence and not an existence, nothing is changed except a false conception for a true one. A reference is made, in our thought and appreciation, to an existence before unnoticed or disregarded. To recognise a shadow is to know the light, to understand that there is more than was supposed, not less. How glorious that being must be, by defect of which is our life with all its beauty, joy, and good; its responsibilities, affections, and pursuits. Even this is not the very fact of life, it is life mingled with death; good enough if it be for us, it is not good enough for God. He who has life in Himself cannot so be content for any of His creatures. For our self-life is not true life, taken even at its best. Let us look into it and see. Is not our being such selves as we are, apart from any perversion or depravity, itself a deception and wrongness? For are we not thus compelled to feel things quite out of any proportion to their true proportionate value? A trifle directly affecting ourselves is felt as more by us, is more to us, than great interests which do not implicate our own welfare. Not because we are bad, but because we are 'selves.' We should be wanting in the essential properties of this state of humanity if we did not feel so. We have to struggle against the feeling; it seems as if we never could be freed from it. By the very nature of our self, we feel things not according to their true value. A less thing is more important to us than truly greater ones. We are not right to the universe. All things are distorted, twisted, turned from their true relations, so far as we are concerned, by the fact of our self-ship. Is there not a fact here which ought not to be ignored? Nay more; often we cannot but feel as good to ourselves the calamities and evils of other men. They are good to the self. And in wrong doing, what is there but the natural result of this warping of our feelings by self? In doing wrong, a man evidently acts against the true value and relations of things, moved by his wrong feeling of them. He violates right to gratify self. Sin is treating things as they are to the self, not as they truly are. Virtue to us means self-denial; yet is virtue nothing but acting according to the truth of things; letting the most weighty have the greatest weight. And when we consider the meaning of the word virtue, that it is manhood, and reflect how it consists in opposing self, surely it should be clear to us that the self is not man's BEING.

Self is our great enemy; it deceives us. It makes us feel things to be good which are not good; evil which are not evil; great which are trivial. If therefore we should be disposed to say, the self of which I am conscious is my true being, or else I am deceived; let us reflect that by this very self we certainly are deceived, and made to feel things as they are not. To God all things must be as they truly are, all felt in right and just proportion. He cannot be a self as we are. He is love, and our self is not love. He is light, our self is darkness.

And again, not understanding that by self a defect is introduced into our feeling, and the action and life of nature made wanting to us, we have been compelled to suppose the most incredible things respecting perception; that we put so much into nature; that it is to us so much more than it is in itself! Light is from mere motion affecting *our* eye, music from mere motion affecting *our* ear; all the value that nature has, it has through the mysterious virtues of our Self, which turns dead mechanical impulses into this various life! Our self a storehouse of all sweet and glorious and wondrous things, to which the universe ministers mere occasion, as it were, to be conscious of its riches! What hard necessity could have brought men to such a thought? But it cannot be true. Not we are rich and nature poor; not so, but we turn out the life from that which is around us, and to our self there is no more the living fact of Being, but inert forces, a mere dead mechanism, which leaves us utterly amazed to think how it can be so much to our perception and our hearts; how glorious we must be to make so much out of so little!

Let us protest one moment against ourselves. Let rational inquiry be heard against assurance. What cause does our experience of nature bespeak? What things should they be which thrill our souls with rapture, penetrate our hearts with such sweet or solemn thoughts, speak with such mystic language, inwind themselves with our deepest feelings, and make themselves part of the very fountain of our life; which are beautiful, gentle, living, full of truth, of majesty, of joy or awe, of comfort or of warning; which are bound to us by infinite relations; which teach us solemn lessons, rousing our souls to ecstacy or anguish: which say to us, Be more, be better, join heart and hand with us?

Matter and dead forces? It cannot be. The theory is self-condemned. It does not account for, nor touch, the most pregnant facts of the case. The question turns itself the other way. We must ask rather why we have been compelled to make such an inference; what that self of ours can be, that has forced upon us such a thought?

For our experience, let us be bold enough to affirm an adapted cause; a possible one. Let us not wantonly make mysteries, and say: Nature is matter and force; but how it can affect us thus passes comprehension: two mysteries; first, that we should know, without a full and reasonable investigation, what nature is, and then that it should be so inconceivably unlike all that its effects and powers would indicate it to be. For there is no man who has not felt in his heart what a miserable failure our investigation of nature is; what poor and even ridiculous results we educe. The secret of its being palpably escapes us: the things we discover cannot be accepted as the facts. Imaginary ideas are invented without end, to satisfy the necessity of finding, in nature, something that at least may seem to make its wonder less incredible. When all that is needed is that we should bethink ourselves, that that which is felt to be by us, may be, nay must be, other than that which IS. Must be; for our self is 'in it'; defect, negation, that which deprives it of reality.

For, that our self is defect of being is, perhaps, in no respect more manifest than in this; that to it the phenomenal is the real. Real to the SELF, unreal to the MAN. Man feels and knows that to be unreal which yet is real to him. The discord of his life is here, in that defect which makes him such a self as he is. The defect, unrecognised to be defect, clothes all things with mystery; surrounds with ever multiplying doubts. The inert and transient world, that does but seem, is the reality to the self. Hereby we know the self: it is that to which the unreal is the real. When we are freed from that, the phenomenal shall be reality to us no more. The eternal fact shall be our reality, that in which is no defect, because no defect shall be any more in us: that spiritual fact, of which our experience testifies so plainly, but which now we cannot find, which disappears when we seek it, because of the blindness in our eye. The life in time is a life of defect, of passion, of getting, from the pressure and torment of our want. The life of heaven, the life eternal, is the life of being, of action, of giving from the riches of our having. Slaves now, rendering a fearful and reluctant service for what we can obtain: Kings then, from royal bounty, of our own freedom, bestowing. No deadly self, coiled like a serpent in our breast, gnawing at our hearts; that we can see nothing, heed nothing, do nothing, but feed for ever its ravenous hunger. Not this self, the Tyrant, the Destroyer. Destroyed by the mightier Love, its pale and wounded victims shall arise, with freed hearts and holy hands, and join the universal life. No more isolated and apart, in self-pursuit, but wholly one with all, rejoicing beyond reach of sorrow, happy beyond touch of harm, blest with God's own blessedness of absolute surrender.

Man is degraded, low, an outcast; open to the cruel shafts of sarcasm, his bleeding heart fair mark for jest and scoffing. Poor is his virtue; a fair pretence, pierced through by gaping wounds, betraying the rotten selfishness beneath. So let it be. Yet stay, O scorner, till the true man be seen. Crushed beneath his enemy, striving vainly with a foe that is himself, bearing death unknown within, and seeking helplessly the secret of his misery everywhere but there; speak not rash words of him. Let death have reverence. Rather, let Life that strives with death and overcomes, the kindling powers that shall know no quenching, the battling light and darkness, the war of heaven with hell, receive the homage of an awful joy.

For He is party to the strife whose Presence is accomplished victory; whom nothing withstands, nothing delays. Who hath abolished death, and with His own blood sealed man's deliverance. Life has been given for life, the law of life has been fulfilled. Christ has borne our death; has brought life into contact with our self and shown it vanquished. The power is there, the work shall surely be accomplished. For He triumphs even now; earth doth confess Him Lord. Suffering and sacrifice, shame and sorrow, by these He is known to be Divine. The weary ears of men listen, and drink in the tale, and own that it is true. Sorrow and sacrifice, God taking our death upon Him, made to be sin for us, these bring salvation, they have a charm which cannot fail. The glad heart bounds to hear it. Herein it reads the secret of its being. The mystery is opened. Heaven is revealed. Now it knows that the self-life is death. It knows God, and lives.

BOOK IV.

OF ETHICS.

He trusted in God.

CHAPTER I.

OF THE FACT OF HUMAN LIFE.

As seeing that which is invisible.

THE views which have been suggested have an evident bearing on the regulation of our life. Our action should correspond to reality and not to a false impression. If therefore man's feeling in respect to the world in which he is be erroneous, it is a necessary consequence, that principles of action based on that feeling should mislead: if we are under illusion, only by escape from that illusion can we hope wisely to conduct our life. Nor is evidence wanting that man does need, for practical purposes, a truer knowledge than he has hitherto brought into exercise. Man fails in his dealings with the world. Not in respect to his action upon phenomena: of these, so far as he knows and obeys their laws, he proves himself the master. He can use the physical world for his purposes, but he fails to conform himself truly to the nature of things: there is a want of accord between himself and Nature, of which he bitterly rues the consequences. Evidently he goes wrong, incurs disappointment, runs into evil while seeking good.

Therefore it were an unquestionable practical gain to recognise the illusion under which we have been in respect to the world, even if we could know no more. From innumerable mistakes it might save us, of treating that as the true reality, which is not;

or of suffering ourselves to think of that as truly good or evil, which we do but feel so. But this is not all. The illusion yields, not to uncertainty, but to a knowledge worthy of the name. The fact and meaning of human experience we know: it is the making man alive. Remembering this, the world stands before us all in light. From an evil state, the worst of states, of which alone it can be said in its true meaning that it is death, the state which constitutes him a self seeker, and is the cause of all sinful deeds, man is being raised: raised to a state of true and actual life, absolute and eternal; death being destroyed, so that he shall have to regard self no more.

The world is being redeemed: this is the fact of human life. Whatever our experience may be, that which causes it is the raising up of man from death. The evidence of this lies in the necessity which has been traced in it. Science, proving that all things are bound together by an inevitable chain, of which no link, even the minutest, might be wanting, gives demonstration of a fact which might overwhelm us with its glory: for thereby we know that all which we experience is necessary for that which is the fact of the world's history; necessary for the carrying out of man's redemption. Christ has interpreted to us the Necessity of nature: has shown us the fact which is the one and universal Cause. For the cause of our experience is not that inert existence we have conceived; the necessity by which all things are determined is not that dead necessity we have supposed. It is the necessity of love: the love of God, which must redeem the world.

Let us say it with reverence: human experience is the making Man alive. This is the fact on which to fix our eye; which, if we would be right to the world, we must regard; seeing which we see it as it truly is. We must remember that man is dead and must be made new; so new that he must be delivered wholly from himself. We are appalled at sin and think: Can that work man's redemption? But it is even so: the evil is not in the sin, but in that which is the cause of sin. By sin our death is made conscious to ourselves, and man learns what he is. If there were no sin there were no less evil, man were not less dead, but the death could not be destroyed. There is no difficulty in seeing rightness, and love, in the existence of sin, while our loathing of it, and feeling of guilt in it, are increased, if we can see the evil of the state from which man is raised; that sinful deeds do but make manifest an evil which exists apart from them, and is their source: an evil which, without them, could not be done away. How evil must be that state of man, which turns the love of God into occasions of iniquity; which, even at the expense of the whole mass of human crime, God must destroy. For the end is worthy: worthy not only of the woe mankind have borne, the tears they have shed, but also of their crimes and guilt. Because that end is not happiness but holiness, the perfection of self-sacrifice, the only good.* It is not the evil that dead humanity should sin: that is a proof and triumph of Eternal Love; for therein man is redeemed.

Fixing our eye upon the fact of man's redemption, all things are made new to us. A glory

---

* Happiness, except as one with self-sacrifice, cannot be said to be good. It is only felt as good, which is entirely different.

almost too great to bear transfigures this poor life: passing all thought and all desire, passing all dreams, and yet no dream, but plain and demonstrable truth. Not less can content the infinite heart of God, nor the Saviour's boundless love: His heart who gives us more than we can ask or think, His love who makes us one with Himself, and bowed His head to death, that He might be the first born of many brethren. For the fact of all our pain, and sorrow, and distress, that for which all are necessary, is the making man alive. God gives to us to suffer for that end; Christ gives to us to be even as He was. He has made known the fact of human life: the Son of Man is revealed in Him. Every suffering, every loss, is borne for man's life, necessary to that end, which could not be without them: necessary, not as conditions or as means, but as the very fact itself, the mode in which it is wrought out.

To believe is salvation: to believe in Christ the Redeemer of the world, that He saves MAN, and will subdue all to Himself: this saves us. It makes us know God; and to know Him is life. To know Him, it is enough. It was ignorance, and blindness, and mistrust, that held us captive, and bound us in the chains of death. Because we knew Him not, nor His work, we grasped at pleasure, we fled from pain and sacrifice, we exalted our own will, and found our good in that which was pleasing to ourselves. To know God turns darkness into light, sorrow to gladness, evil and pain and wretchedness into unbounded joy: makes welcome to us every loss, ennobles all things trivial, gives to earth the blessedness of heaven. To the saddest, weariest hours comes a delight which makes all other pleasures poor. From the sorrow the rejoicing springs: we are suffering that man may live. Therein are we one with Christ, whom all hearts do pronounce Most Blessed, and fill up that which is behind of the sufferings He bore. We too are ready to be offered. Our hearts are taken captive utterly by love. The terrors that have haunted us, the evils we have shunned, were but dark shadows from the blackness in ourselves. We look abroad again, and the light of heaven glows unchequered over all. Our fears are gone. If there be no evil but that which love makes necessary, then there is no evil: no pain but pain borne for man's life, then is pain utterly transformed. The one Love, that is in and through all things, by which all things are, the Love that is the only joy, smiles also through the tears of sorrow. Life stands confessed beneath the mask of death.

CHAPTER II.

OF ILLUSION.

From the ingrained fashion
Of this earthly nature
Which mars thy creature
From grief which is but passion;
Good Lord deliver us!

THAT man is being raised from a dead state into life gives a solution of the otherwise insoluble problems of our experience. The light which this thought throws upon sin, showing it to be a necessary result of the death from which man has to be delivered (the evil of which it displays and brings to our consciousness, but cannot increase), is but one example of its bearing upon the mysteries that throng around us. Our perplexity arises from our inadequate apprehension of the fact that our present state of being is not the LIFE of man; that ours is not the true humanity. For, thereby, we tacitly take ourselves always as the standard, and assuming in our thoughts that which we feel as that which IS, we surround ourselves with darkness; not reflecting that a being in a wrong state must feel, as well as act, wrongly. Thus, for example, has come an utter confusion into our ideas respecting good and evil. For it is evident that the raising man from death to life is good, the only possible true good for him; and whatever things are included in, or necessary for, that end must be also perfectly good. But these things are felt as evil by us; to us they are evil, involving the loss of that which we value, the failure of that which we attempt, the bearing of that which is painful. We feel as evil that which is good. It is evident that we do so, for we feel as evil that which God does: things which constitute part of the course of nature, and cannot be dissociated from His immediate agency. The fact, therefore, with which we have truly to deal is, that we feel as evil things that are not evil. We should ask: What is the matter with us that we feel God's act as evil? But instead of this, taking our feelings as expressing the truth, we have been asking: How can there BE so much evil? By our self-belief, and confidence in our own impressions, our whole thought has been perverted. Why must we feel so much evil? is quite a different question from the other, How can so much evil be? and especially different in this, that it can be answered. We must feel evil because man is in a wrong state, and is to be made right. It is not an evil thing that man, being as he is, should feel evil; it is not a mystery. If his feeling were not of evil, the fact could not be good; for he does not feel things as they are. Nor can the amount of the evil felt by him, however absolute or enormous, affect the case. It does but prove more emphatically the wrongness of his feeling, and place in clearer light, thereby, the wrongness of his being. That cannot be the true humanity, by which things are felt as they are not; God's act felt as evil. The false feeling proves the wrongness of the state to which it belongs.

Evil pertains to the phenomenon. The feeling of evil is inseparable from the feeling of the phenomenal as real; it is inseparable, therefore, from the present state of man, the essential character of which is that phenomena are real to him. Evil, therefore, must be felt as real by us, until the defect of our being is done away: but this false feeling on our part need not deceive and mislead us: we need not act as if it were true.* Even while our feeling continues erroneous, our belief and understanding may be right, and our action thereby be redeemed from misdirection. We can recognise the cause, in our own condition, which makes us feel as we do, we can fix our eye upon the fact, we can make our action true to that which IS. In all our feeling of evil we can think: I am feeling as evil that which is not so. We can make even our words conform to truth, and say of every misfortune, not 'how evil this is,' but, 'how evil I feel this.' Nay, we may go further, knowing the fact of man's redemption. Even the feeling of evil, though it cannot be altered, may be swallowed up and lost, so absorbed into a larger happiness that its very character is changed. If through the feeling of evil man must be redeemed, how could we be willing not to bear our part? Is it evil that we should feel pain, should suffer loss, that man may live? Can we fail to find in that our truest

* However much evil there might be in these things, it is clear that it would not be 'real' to a being to whom these things were not real, to whom they were but appearances: the evil to him would be apparent only. This is the truth: evil is apparent, not real. But it is felt as real by those who feel apparent things as real. It is real to those in whom there is defect; in whom there is such a self as ours. To be delivered from evil we must be delivered from our self. Then, when the fact is truly known and felt, evil is no more real: it is known and felt as love.

blessedness? To regard the fact; to know the redemption of the world, and to fix our eye on that; this is the cure for sorrow. This is the gospel, the good tidings of great joy, which are to all people, flooding earth with a sea of gladness.

Christ shows us that the world is going right; not wrong as we suppose: that it IS full of God's glory. So He gives us happiness. For the sting of evil is that we think it truly evil; that we do not know that the fact is good, utterly and entirely good. If we will recognise the death of man, his need to be made alive, and the fact that he is being made alive, and that all human experience is necessary for that end, evil is no more evil. All this is the form in which man's redemption is presented to us, and we therein rejoice and will rejoice. What does it matter that it is painful to us, if the world therein be saved?

The mystery of evil is that we feel it, not that it is: and that is no mystery. Christ has taken it away, showing us the death of man that is, his life that shall be: showing us the meaning of all pain, and loss, and failure; that it is for the life of man. What we naturally seek and desire is to have the world as we should like it, conformed to our feelings of good and right; such as suits us in this present state. And we think it evil that it will not be so, and lament over the mystery of God's ways. But to have that which we naturally desire, were to have man confirmed in death, were to forego his redemption. He must be made different from that which he is; therefore the world must be evil to him. Because it is too good for him, he feels it evil; because it is so truly, absolutely good. We would have it good to the self, but the only goodness is the destruction of the self. For which destruction, misery must be, and cruellest pains, and crushing of tenderest hope and love; nor only so, but rampant guilt, and wrong triumphant, and sacrifice of noblest aims. That which were good enough for man's desires is not good enough for God. If there were not that which we feel as evil, that perfectly good fact which God wills could not be. In all afflictions we may say: it is very hard to bear, but the world is going right: shall I not bear this, if man be saved and this be necessary? If this cup may not pass away, shall I not drink it?

Is not Christ's death the joy and glory of the world, the best thing that ever happened? In it we may see the difference of the fact from the appearance. For from a human point of view what is the death of Christ, but a black murder; deceit and violence crushing the hopefullest of lives? 'We thought it had been he, who should have restored the kingdom unto Israel.' But more was to be done than that, more and better; a work for which not His life only, nor His agony, but lives unnumbered must be given, and agonies drawn out through every age. At all expense of woe, man must be made alive.

Thus we see, also, why man's present life must have been such a deception: why he feels that which is phenomenal as reality, finds the spiritual physical, and has been condemned all these ages to believe it so. To have been in error and deceived was right for him. This also is the fact of his redemption; for so he pursues unrealities, places his confidence in that which falsifies its promise, seeks rest where no rest can be found. So he fails, and learns his ignorance; so he rushes into sin, and learns his vileness. So he is taught that in his very being he must be made new; must be delivered from himself, and be made living with an eternal life. The reason of man's perceiving as he does, feeling himself in a material and transient world, is that this experience is necessary for the work of his redemption. He must have this consciousness, that thereby his state may be altered: he is under illusion, not that he may continue so, but that he may be delivered; that he may feel its evil, and escape its tyranny. 'The creature is made subject to vanity,' but it is in hope. We cannot be holden of its chains. For what bondage is so wretched, what slavery so degraded, as being ruled and driven by an illusion; spell-bound beneath the power of a phantom world? To know that we have been so deluded is itself deliverance. Once convinced that the eternal things are the true realities, our slavery is at an end. Man wakes from his troubled dream to see the glory of eternal Day around him; braces himself to waking life, and looks back, with mingled gladness and surprise, upon the dim chimeras which his unnatural slumber had invested with a brief reality, a transient power to make him glad or sorry, hopeful or afraid.

CHAPTER III.

OF REALITY.

The foulest act with which man's hands are soiled,
That telleth, loud, humanity's disgrace,
Leaves on the earth its evil; undefiled
The Fact uplifts to heaven its holy face,
And blotteth not the pages of that book
Whereon the brooding eye of God doth look.

THE error of our feeling is that phenomena are the reality to us; the error of our practice is that we treat them as reality. We do not see the worth of temporal events. Overlooking their relation to the fact of man's redemption, we overlook all that is real and absolute in them, all that constitutes their value and necessity. If in all things we regard the fact, letting our thoughts go on, beyond our own impressions, to that which truly is, then we can treat all things aright. For the world's redemption is a fact which, by its nature, surpasses and subordinates in our regard all others. In its presence, illusions lose their power; new forces influence us; the world is not to us what it was before, but infinitely more. Our whole being is enlarged. A new and overpowering thought absorbs the private regard. It cannot grieve us any more to suffer, or to forego our own desires. A joy so great springs out of the suffering, in our knowledge of that for which it is, that suffering itself is changed. Knowledge of man's redemption, which shows all suffering to be suffering for the world, makes a new thing of human life; inverts it; more than doubles it; extracts from that part of it (how large a part!) which we have deemed mere loss and evil, a value infinitely exceeding all the rest; makes suffering more to be desired than that to which we have heretofore abused the name of joy. For in suffering we are one with Christ. With what a radiance it crowns anew the brow of Jesus; making the sacred words: 'I lay down my life for the world,' more sacred evermore.

We have not seen the truth of human life, we have overlooked the very fact of that which it IS; what wonder then that it has been found a gloomy mystery? for that men have found it so cannot be denied. It stands written in imperishable records. Literature is man's thought of life, wherein he gives verdict that it is inscrutable and dark. And it is so utterly; unless man be in a state which makes all the evil that attends his course necessary for his deliverance; unless we can turn the clouds that are around his path to brightness, recognising their true secret; and understand that this Being, so surrounded and penetrated by evil, is not the true Man; this monstrous course of crimes and errors not man's life, but the making him alive. The evil cannot be denied: it is too foul, too loathsome. The universal conscience of mankind rises up to rebuke him who would make light of it. To interpret human life we must learn some unknown fact, which shall bring evil into a new light. This unknown fact is supplied by the perception of man's death. This takes away all evil; or leaves it, rather, but as the manifestation of infinite and boundless love, the Perfect Good. Our thought that things are really as we feel, our belief that man has life, alone prevent our knowing and feeling the truth, and compel us to unite evil with our thought of God.

We cannot be happy, nor content (it were inhuman to be so), while we think that there is truly evil, that the world is going, even in part, to an evil end. And we cannot, without the grossest closing of our eyes to facts, think otherwise, unless we can recognise something more in human life than it appears to us; a reference to some other end than those which we naturally suppose. For in respect to those ends it is a failure palpable and manifest. Neither man's enjoyment, nor his virtue, is secured. But to know man's need of redemption from a state of death, and to understand the fact of it, makes all things right; gives us a source of happiness perfect and unassailable, a spring of energy which nothing can damp. Be the phenomenon what it may, of agonising misery, or wrong unspeakable, our eye is fixed upon the deep underlying fact: in this, too, man is being delivered from himself, being raised from self-seeking to self-sacrifice. Nothing could be spared, nothing could be otherwise. Man, being as he is, must so be made alive. Nothing is evil in that which IS. In the saving of man from self-dominion, from wrong-feeling, and the possibility of sin, there is no evil; none in the fact on which God looks, and all look who truly see: all who see aright the state in which man is, and the Necessity that he be saved from it.

Seeing thus in all things the fact of man's redemption, the world is made a new and different world to us. The things which make us do evil lose their power; we no more covet, are no more afraid. Our action becomes right to nature; conformed to the truth of things; it cannot end in disappointment therefore; cannot fail. For we feel truly. In all things we look at man's redemption, and when it comes that we are made to suffer and to lose, and our desires are set at nought and frustrated, then glows our heart with a joy unparalleled, too great for words, filling our faith and love to the uttermost, making us know what heaven is, and what the joy God chooses for Himself, the eternal joy, wherewith the infinite fulness of His bosom throbs, whereby He is THE BLESSED God. His own joy, but a joy He keeps not to Himself, but gives to us also, too unworthy; the joy of sacrifice for man's redemption. Passing belief, and yet not passing; the humblest faith must stretch her hands even to that height of glory, and human love expand its puny measure to become the heir of God. Of God revealed in Christ, made manifest in Him so that we know Him, truly actually know Him, and see Him as He is. He IS the Sacrificer, joining us with Himself therein: that is to KNOW GOD. It is to be one with Christ the Saviour. We can believe it, for though it is we receive the too great blessedness, it is God who gives. Who is the Giver, whose happiness is in giving, all whose gifts surpass our thoughts and fill us with an infinite surprise.

Oh weary and woe-stricken world, oh vainly striving men, your misery is that you do not know; that you see not that which IS:—Why you are sacrificed, why you are wretched, by what NECESSITY such cruel pains assault you, such bitter lacerations of the dearest ties. A sevenfold mystery besets you round, and tears as of blood drop from your darkened eyes. Behold and see! The Man who has gathered up all sorrow, and deprived it of its sting: the pattern of all mourners; the revealer of the Father. Giving His life for the world He stands, the head and crown of humanity, which drinks in life from Him, and grows into His image. We too are made conformable to His death, drink of His cup, with His baptism are baptized. For if in our agony the world is saved, ours shall be a willing agony like His.

There is no true consolation for sorrow but in knowing man redeemed; in knowing that the world is going right. That is, in fixing our regard upon a good so great, so much better than anything we naturally regard, that it makes all evil good, and satisfies us for the loss and overthrow of our best desires and hopes. Placing our happiness, not in having what we like, in obtaining what we should choose, but in the goodness of the FACT, the only rightful ground of happiness, we have attained a perfect consolation in all ill. Revealing the perfectly good fact, the redemption of man from being such as he is, Christ gives peace: a comfort not infected with the bane of self-regard. For only by knowing and believing this, can we escape that worst of wretchedness. If man be not going right, if all be not well for him in spite of seeming, in what can we rejoice but in the thought that we are or shall be happy, though others be not so? which is the greatest of all curses, the very depth of death. All that is in us, not utterly corrupt, abhors that most of all. Therefore do some men deny that man is corrupt, and represent sin as a trifling thing, repudiating God's just anger and the inevitable punishment. They crush the voice of conscience, because they cannot be content that Man should not be saved; because they will not place their happiness in their own unparticipated good. Therefore do others say: our happiness will be in God's glory and perfect justice, irrespective of the doom of those who reap the just reward of guilt; crushing the human instincts in their breast by the overpowering force of conscience. For what man could be happy in his brother's execution, because justice is fulfilled, and the honour of the law, God's law though it be, maintained? If this be man's life, and therewith his probation, the conscience and the heart are hopelessly at variance. But see the reconcilement of this strife in the recognition of man's death and his deliverance. Man's redemption from a state in which he wants that which constitutes his Life, known as the history of the world, fulfils all the demands of our nature, justifies all our convictions, satisfies all our aspirations: accounts for all that is in us, all that is around: displays the dark and tangled labyrinth in clear and glorious light.

For even such a deliverance we need: to be saved wholly from ourselves. When we look into our hearts, how mean and unworthy is even that which is best in us, how utterly marred and spoilt. A self-regard lurks in it all, which we would fain hide even from our own eye. And in this we feel it most, that when we have done the best deed, and most have given up ourselves, then the hateful thing will come again, and we think: Now I have done well, then I was truly good. 'Who shall deliver us from the body of this death?'

Christ shall deliver us, by the Fact He shows us, the God He makes us know. By showing us what Life is and what Death, and what all things are, making the world so different, He shall drive that demon wholly from our hearts, and God Himself shall dwell therein. He shall save us utterly from self-regard, swallowing it up in an absolute and boundless joy in that which IS. A joy in God, perfect and complete, a delight so full in the eternal good, that no self concern can have any place: self sacrifice so perfect that it is no more a sacrifice.

For his true salvation, man needs to be delivered from himself, from the necessity of making his own gratification his chief object. The problem of humanity is to make love the spring of human actions. It is apparently a hopeless task: experience seems to forbid the expectation, reason will not sanction it, calm investigation of what man is shatters the fond dream. It is man's nature to act from interest and from passion. It seems impossible to do more than wisely to direct his regard to his own welfare, that he may prefer that which is most to be valued, and subordinate the passing and inferior to the superior and endless. But this is the question whether the dead can live. If the true fact be known, it may be answered the other way. Believing man's redemption, and knowing what the world truly is, love must take the place of self regard. It asserts its rightful power. Its lost dominion is restored. With that alteration of our thought, our life too is altered; we cannot be as we were before. The whole scope of our regard, affection, and desire is new. We are taken out of ourselves by a Power above ourselves, and understand what that saying means: 'That which is impossible with man is possible with God.' We want no more, because we have. There is nothing for us to desire, except to be made free from the remaining death: no other blessedness, no other glory than this we have, of being one with Christ: only that we might be fully one with Him.

And the glory of this Life embraces not only the great events which rouse enthusiasm, and kindle energy in all; it extends to the mean and ordinary incidents in which we are so weak. There is no great or small, where all alike is necessary. Nothing is so trivial, that in it the eternal fact is not; nothing so mere an accident that it must not have been for man's redemption.

CHAPTER IV.

OF WRONGNESS.

Either make the tree good and his fruit good, or else the tree corrupt and his fruit corrupt.

Let none admire
That riches grow in hell, that soil may best
Reserve the precious bane. *Par. Lost.*

USING the world as we naturally do, we can attain certain ends and achieve much that we feel desirable. Yet an incurable fatality seems to attend all our actions. Permanent satisfaction fails us. The results will not answer to our hopes, and a mysterious necessity of evil seems to be in the world, that leaves the most sanguine, at last, hopeless of a remedy. This experience is too familiar, and has been described too often, to need any more to be insisted on. But there is an advantage in understanding it, and in seeing that it is no mystery, but a natural and necessary thing. It is the inevitable effect of error. Whoever attempts any work under a false conception, meets with the same experience, as men find in dealing with the world as if it were, in reality, what it seems to them. Under a misapprehension of the true nature of that with which he has to do, a man places certain objects before himself which he feels convinced will ensure what he desires; he uses means in pursuit of these objects with ardour and satisfaction; he rejoices in them, he is assured he will succeed, he seems to himself to have all that is needed for success:—but the end is failure. He is astonished, perplexed, angry; he feels sure there has been some accident, perhaps some precaution omitted, he repeats his efforts. All the old enthusiasm is rekindled for a time; there is the same ardour, and satisfaction in the work, followed by the same disappointment and recoil. It must be so: his energies are misdirected. He is acting under a false conception, according to appearance, not according to the truth.

Just in such case are we. In dealing with the world as we naturally do, we are treating it according to its appearance, and not according to the truth. We are acting under a false conception. We have enthusiasm, and pleasure, a certain satisfaction in our pursuits, and a firm expectation of success:—But what is the result?

For true success there must be right knowledge. We are not acting according to the truth of things.

While men conceived the sky as revolving round the earth, they could learn much respecting the appearances it presented, could ascertain many laws, and by means of that knowledge accomplish many things; predict eclipses, and conduct short voyages out of sight of land. But for the true uses of astronomy, it is necessary to recognise the cause of the appearance. Fixing their thoughts upon the true relations of the universe and the earth, men have a power, and effect results, which else were unattainable. And for the true uses of the world we must understand it rightly. To recognise the cause of the appearance, and to fix our thought on the fact of man's redemption, is to act by the world as being spiritual, to treat it as it is.

Again: Not only does error necessarily lead to failure, but, in this very failure is the remedy for the error. We learn by failing. Starting from a state of ignorance, we necessarily act upon false conceptions before we obtain true ones, and are delivered from our ignorance only through the evils which it brings upon us. All human history tells the same tale: of wisdom learnt through error; defects remedied through loss. The practical evils of the outward life make us aware of the defect within. Save by passing through mistakes and failures, which make him feel himself wrong, in apprehension or in feeling, there is not true good for man. The evils which he inflicts upon himself teach him the lessons most necessary for him to know, but which without them he could never learn. They are necessary, therefore, and good in their evil. Their goodness is in their felt evilness. Man's errors are rightly wrong; their wrongness is their rightness. They are wrong for the results he aims at; right for their true result, of making him wiser, and curing his defects.

Even so is man in his relation to the eternal. His defect is made conscious to him by its results. He learns through error. By his failure he is forced to recognise his misapprehension, and to know that the world which he treats as physical will not be so treated unavenged. Truly the world is wrong; but it is rightly wrong. Therein the true ends which it subserves for man are perfectly fulfilled; though not the ends which he proposes. The phenomenon is wrong, but the wrongness of the phenomenon is the rightness of the fact. If the world were right to man, his deadness never could become known to him, never be removed. This wrongness and evil are the removal of his deadness; even as his errors respecting material things are the removal of his ignorance.

The work of making man alive, and truly Man, we feel as evil. It is evil to the self of which we are conscious. But so it ought to be. The work of making man alive is a larger good than our capacity can grasp: it includes ourselves, involves our being made different from that which we are. Therefore to us it must involve the feeling of evil. That is no more than that a child should find the arrangements of its home bring with them that which it feels as evil; that being often felt by it most evil, upon which its welfare most depends.

Man must have learnt the truth respecting himself and the world through error, and failure, and sin. Only so is ignorance corrected, only so can death be done away. But in other points of view, also, it may be seen that the world is right in its wrongness, and in its felt evil. Its very nature and constitution involve its evilness to us. Our feeling of good depends on that of evil, pleasure involves pain. Without the consciousness of the one we could not know the other. To do away with suffering, and leave that which we call enjoyment, were impossible. The feeling of evil is a chief stimulus of all our energies, lends its vigour to life, and plays a main part in the raising man from barbarism to civilization. It could not be foregone without the utter ruin of social life. Man lives by his wants. In truth, we might well believe that upon pain and want this physical consciousness is founded; that the fundamental idea and basis of this life is pain, and not pleasure, the latter springing wholly out of the former, and implying for its possibility the previous existence of discomfort.

But whether this be so or not, it is certain that where there is a self, such as that of which we are conscious, there the feeling of evil, or at least the liability to it, must be. The self carries the consciousness of evil with it. We cannot conceive self apart from the wish to avoid and to get. Above all, there must be the feeling of evil where there is the self, or how could there be self-sacrifice? And how could self-sacrifice be foregone; the one joy and beauty and delight, that does not mock the name; the one thing that redeems the earth and makes it worthy of its place in heaven? How can there be self, and self-sacrifice be impossible, until Satan has triumphed over God? Yet this must be, if we separate from self the feeling of evil. If nothing painful were to be encountered, nothing to be sacrificed, how could there be at once self and love?*

But the question will naturally arise: if we are to think thus of the things which constitute the realities of this earthly life, will not this lead us to neglect them, and let them fall into disorder?

---

* Love necessarily involves that which is evil to the self. Love is in sacrifice. But where love is, there the evil to the self is no more felt as evil. Self-sacrifice alone is joy. This is all man wants to put him in heaven. Perfect love, that the evil to the self be no more felt as evil, but converted into joy. He wants LIFE that he may be in the eternal world; he wants the self destroyed. For is it not strange to think what lurks in our thought of heaven? This self shall be in us still, but there shall be no possibility of self sacrifice: there shall be nothing to be suffered. This is why the devoutest men can hardly find food for enthusiasm in the thought of heaven. They ought not. That heaven is not so good as earth; where with all our shortcomings we can still sacrifice ourselves; can still endure for love.

Why should we attend to our business, why seek to advance the comfort or maintain the order of life, if the goodness of the world be in its evil? A little reflection will show that a regard to the fact, instead of the appearance, would be as advantageous for the comfort and progress of the earthly life, as for higher objects. For, whence come the disorders and evils of that life, but from self-seeking, and from too intense an eagerness to possess that which is pleasant? How could evil result from substituting love and duty as the prompters to activity, instead of ambition and desire? That were surely an excess of caution, which should dread the effects of too great a subordination of the self-regard, to joy in the work of God. What evil so afflicts us now, as it seems to some, above all other ages, as the frantic desperation of all men to do well for themselves? Or in what have the moralists and wisest men of all ages so agreed, as in testifying to the want of something which should moderate the violence of the passions? Do we not, also, well know that incomparably less toil, freed from the perverting power of selfishness, would ensure a far more rapid progress, than is ever likely to arise from the conflict of private interests?

Is not virtue the true wisdom? Does not the truest worldly success attend it in the end? Do not crooked courses lead to loss and ruin? Yet what is virtue, but the treating these things as of no value in themselves; scorning them, casting them utterly aside, as merest trifles, in obedience to the fact which speaks in duty? The glory of virtue and nobleness is that they treat the phenomena as phenomena: they are true to the nature of things: they are success. And all meanness, vice, and hatefulness, what are they, but the treating these things that we like and fear with an undue regard, as if they were the realities?

Or again: There are two agencies which prompt us to action; our desires and our conscience: the stimulus of pursuing that which we enjoy, or escaping that which we feel painful, on the one hand, and the sense of right upon the other. Let it be conceived that the former of these were done entirely away, that we were made absolutely indifferent to pain or pleasure, what then would regulate our actions? What would remain? Clearly the conscience. The sense of right alone would regulate our conduct. Would the world be worse or better, if men were moved to action only by the conscience? If it be asked; what should we do, if we believed that all the things we feel as evil were truly good? the answer is evident; we should do right. We must do so. The belief in Christ as the Redeemer of the world saves us from sin.

But in truth, this doctrine of the world, so far from diverting attention from the practical matters of our daily life, is of all doctrines the most practical. It exalts these things to their true dignity, raises them from a false and most fatal depreciation. It affirms for them a value not less, but infinitely greater, than that which is assigned to them; but it regards the fact of them, and not the seeming. These things are not merely the trivial things which we suppose; they are the mode in which the eternal presents itself to us. We slight them and do them wrong, give them not half the heed which they demand. These are the very facts of the eternal world, besides which there are none other. These are the facts with which our concern lies—our sole concern; the fact of these daily, ordinary duties. The present is the Eternal. Vainly do we fancy to ourselves an eternal in the future. The eternal is now, or it is never. The spiritual world, and the material, come not in succession even to us; they co-exist. They are fact and phenomenon. The material is the appearance of the spiritual. Why we misuse the world so much, is that we estimate it too low. We do not see enough in it, therefore we so abuse it. More rash and reckless far, than he who should use gold for brass, we squander Heaven's own wealth, as if 'twere merely gold.

There is a fatal practical defect in the belief that the physical is one FACT and the spiritual another, which may go far to account for the apparently incurable errors of our lives. Disjoining thus the one reality, we can rightly apprehend neither world, the apparent nor the true, still less rightly act by them. Either we regard the present as merely physical, a matter for enjoyment, for doing the best for ourselves; or, if we seek to regard the spiritual, it is as another different world, drawing our thoughts away from that which is now around us, and mostly, as pertaining to the future. So we vibrate between a worldly regard to this world, and a spiritual disregard to it; with what results we see. There is a strife between our religious and our earthly life. But the true regard to the world is a spiritual regard to it; a regard to the fact. Not two things, but one, are the religious and the earthly life: the one the fact, the other the form; answering to the true relation of the eternal and the temporal. So grows our life into one harmonious whole; the living fountain within springing up, and filling the else empty cisterns of this life of forms. Not foreign to our piety nor deadening it, but its very life and being, are the tasks of our earthly course, the routine of our daily work. Seeing these as they are, regarding the fact of them in man's redemption, they separate us not from God, but draw us to Him; they bind us to Christ, whose life and death alone enable us to understand them, fill them with all their meaning, make them to be that which they are to us. To us to live is Christ.

But, as we naturally think of the world, not only do our passions and our self-seeking pervert our actions and draw us aside from right; even our best impulses and desires lead us astray. Nothing has been more productive of mischief than ill-directed zeal for good. We snatch prematurely at results, impatient of error and delay; and so we mar the working of beneficent laws, and fall into errors which can be redeemed only through the most terrible convulsions. This comes from our want of faith, from not seeing that God IS glorified, and that His glory can suffer no diminution. On us it can depend, only, whether God shall be glorified in our willing action. Not by securing certain results rather than others, but by simple right doing in spite of all consequences, alone, can we subserve His glory. He does reign, and His kingdom ruleth over all. The perfectness of Eternal Love is here and now. No care of ours can take that which we feel as evil out of the world, nor ought to do so. Man's salvation is in it. Thus seeing, we can keep our action straight, level to the true line of uprightness, and are delivered from the evils which flow from attempting to do good unrighteously. For we leave off acting for results. Results belong to God. We know not what is good, even in the narrowest sense of securing our own greatest pleasure; much less what is good for man. For the true goodness of this world is neither enjoyment, nor virtue, nor any other thing that we call good, but that which IS: man being made alive from death, and raised to a new BEING.

BOOK V.

DIALOGUES.

Hylas.—You set out upon the same principles that Academics, Cartesians, and the like sects usually do; and for a long time it looked as if you were advancing their philosophical scepticism; but in the end your conclusions are directly opposite to theirs.

Philonous.—You see the water of yonder fountain, how it is forced upwards to a certain height; at which it breaks and falls back into the basin from whence it rose: its ascent as well as descent proceeding from the same uniform law or principle of gravitation. Just so, the same principles which at first view lead to scepticism, pursued to a certain point, bring men back to common sense.

The following Dialogues are expository, not controversial. They do not profess to answer all objections to the views that have been proposed, but are designed rather to exhibit them in relation with a wider circle of thought. To a large extent, arguments already suggested are urged under fresh aspects, and with the view of guarding against misapprehension the same essential conception is presented in varied modes of expression and illustration. Each of the Dialogues, however, embraces subjects not previously treated.

DIALOGUE I.

READER AND WRITER.

R. IF I have understood you rightly, what you say rests upon this principle: That the defective state of man causes our feeling not to correspond with the truth of things; so that we can only understand aright either ourselves or the world by remembering that man is wanting in life.

W. It is so. I say that all defect, perceived as absolutely existing, apart from us, proves itself by its very nature to be due to man's own condition; implies defect in relation to him.

R. Your position, I grant, is a reasonable one to consider; but there remain many grave objections. I will not mention the strangeness of the idea, and the alteration it demands in our way of thinking. That may be due only to its novelty. It may be as natural to conceive of defect within us, as without us—of ourselves as being conscious of defect, as to conceive the opposite—when once we are familiar with the thought. It would be unfair to press you with that as an argument, which may rest only upon custom. But let me mention, first, an objection which should weigh much with every reasonable man. Do you not put yourself in opposition to the universal opinions of mankind, and give direct contradiction to sentiments which have all the authority that human conviction can bestow?

And this, not on some few points in which we might expect that error should be detected, but in relation to the entire scope of human thought. Is it not most unlikely that you are right?

W. If the case were as you have stated, I should agree with you. I should think any deductions which one man might make, however supported by argument, to be of little value, if they were in opposition to the real convictions of mankind. Understand me better. All my confidence is placed on the very ground on which you would have it rest. If I have not uttered the true convictions of man, and have not on my side the affirmations of all who are most worthy to be believed, I would wish everything unsaid. I am a learner, not a teacher. All that may seem new I have learned from the living lips of men, or from pages on which man's life throbs inextinguishably. I have not said one thing that you may not find better said before, or gather, fresh with the dew of simplicity, from the way-side as you walk. Mine has been an humble task: to listen what men say, and let it sink into my heart, and repeat it to themselves: the child's part, who may sometimes see what his elders overlook, because of his conscious ignorance. For men, sometimes, in their great progress, seeing so many things, have not time to attend to all, and may suffer old ideas to remain in their minds, without observing that they are incompatible with their better knowledge. Listening to the large discourse of humanity, with an humble heart, and willing to be taught, as became one so little worthy to do more than the scholar's part, I have heard it affirm all the things that I have said. I have heard and believed, for they seemed to me true; and I could not help seeing that they formed a consistent whole, and that men need not contradict one another any more. For who does not, in his heart, affirm nature to be living and active; when has it been otherwise spoken of by man, speaking his true thought? And in what age of the world has not a deadness been recognised in man? who is not conscious of the sad truth in his very soul? But that there is a deadness in man and that nature is not dead, is all that I have said. It is not I that say it, but man. These two truths had not been brought into relation. For if there is a deadness in ourselves, how could we but perceive a deadness in nature, and become conscious of it, to our wonder and distress, when science taught us to observe? And who does not say that we are in the eternal world; that God and all spiritual being, if there be any such, is here and now present; and that these things are only hidden from us by our inability to see? the present state of man making them to be to us as if they were not. I bow to the assertion: it is true. We are in the eternal world; the very actual world in which we are, that is the eternal. And when I hear the men of science say that all the things which sense and thought present to us are but phenomena, and that the very fact of being is unknown, how can I help recognising here that which I have assented to before? These things must be but phenomena, for they are not the eternal, they are not that world in which we ARE, they are the world in which we seem, and feel ourselves, to be. And when, again, I hear it said that these phenomena are the realities of our life, the only things with which we have to do; when it is affirmed that these things, which are not the very fact of being, are the facts to us; how can I help recalling what

I had been taught before, that there was a deadness in man? how could I help seeing that a life in that which is not the very fact of being is not the very fact of life?

And I have read no book to which man has confirmed the meed of immortality, I have conversed with no pure and truthful heart, that did not affirm to me the unreality of earthly things; that these things, which are real to me, strictly are not, but are shows and forms, which to trust is to be deceived. How could I help seeing that I needed to be more; that the things which are but shows might be but shows to me, the things that are truly real be the realities? How could I not perceive that things ought to be known and felt by me as they are, and not as they are not? And when I heard almost all men affirm another being for man, in which these things should be to him no more as they are now, but he should be in the eternal world; how could I fail to see in that a deliverance from the deadness, a life to that which IS, phenomena being no more realities? 'Accuse me not of arrogance.' How could I believe the true things that men say, if I believed not as I do?

R. You would imply, that in these things the true utterances of man are found, and that the statements which are inconsistent with these are only inconsistencies, not contradictions.

W. Even so. The apparent contradictions are seen not to be truly such, when the hurry and excitement of our life subside, and we have time quietly to look into our thoughts, and see what we really know. For men truly know much more than they are aware of, if they would only bring the knowledge, which lies scattered in isolated portions, into its right relations. Opinions which seem opposed often need only to be regarded in connexion, to give to each other mutual explanation and support.

R. For instance, that the world is physical, or consists of inert matter, means only that it is so to our feeling or consciousness. We perceive it so, and have been obliged to draw corresponding inferences; which is an essential part of your representation?

W. Yes. But one chief test of an opinion is that it should embrace, and draw into itself, all the opposing views, and show each one to have been necessary in its place and order; so presenting the history of human thought as a true living development. What, for instance, so confirms the Copernican interpretation of the heavens, as its explaining the order of men's thoughts respecting astronomy?

R. I grant that if, by a condition affecting man, our perception of the world is modified, and caused to be of an inert instead of an active existence, men must have constructed science as it is, and ought to find such an interpretation of it as you suggest. If the absolute being in nature be spiritual, and man be defective, doubtless investigation ought to make him aware of those facts by such a process as you say. But this brings me to another remark, which is not so much an objection to the truth of your idea, as to its value. What claim to certainty can any such speculations possess? Innumerable solutions of these problems have been put forth, each one announcing itself as successful, but all, as you necessarily imply, erroneous. Why should the last have any better fate than its predecessors? Is not the prudent caution, with which all such attempts are now regarded rather as interesting amusements than as serious work, justified, and as much so in your case as in others?

W. Far be it from me to attempt to exalt myself by a depreciation of others; any speculations of mine are of no more value than the idlest of the past. Nor do I enter into any competition with those men of gigantic ability, who have reared speculation to a height which has demonstrated to all future time at once her power and her incapacity. If I claim for my work a more permanent value, it is precisely because it is of so much humbler pretensions. I present to you no speculation, no attempt to erect man's intellect into a judge of the universe. I present to you, indeed, nothing of my own. But this I say: that man is the INTERPRETER of Nature as well as her servant, and that by science he has interpreted her. It would be a false modesty which should prevent my insisting on the value of the work that MAN has done. What part have I in it that I should pretend it to be less than it is? I say, that science demonstrates that the perceived inertness is due to man: demonstrates it as certainly, more certainly (if demonstration admitted of degrees), as that the motion perceived in the heavens is the earth's. If any man will say that there *exists* an absolute negation, I will allow that, to him, I cannot prove my fundamental position, and consent that the doubt shall remain; but even then, I will not admit it more doubtful than all things else must be *to him*.

R. By science, you would say, the conditions of the problem are altered, and that which was impossible made possible. If the inertness must be ascribed to man, the statement that he has not truly life is but another mode of expressing the same thing. You would make the entire position a question of science, not of speculation.

W. That is what I would do. I only ask the question, Science answers it. Thus would be attained that certainty and demonstrativeness in philosophy, which has been so earnestly sought, and latterly pronounced so hopeless. All men agree in scientific truth, bowing to evidence not to be questioned. Why should not all men agree that the perceived inertness is man's? It is a question of science. It cannot remain in doubt; it must be decided one way or the other.

R. That question certainly alters somewhat the scope of philosophic inquiry. If the inertness be demonstrably due to man, something at least is done; but can this be *demonstrated?* is it not only an opinion still, although we may grant that the opposite may be reduced to a verbal contradiction?

W. I have tried to give demonstration of it; but the question must rest with each man's thoughts. I would rather ask than answer it. I cannot doubt what any man will reply, who will ask himself: I cannot expect or wish that any man should suffer me to make reply for him. Try yourself to conceive the case; the inaction of nature, as it is to us, is *absolute:*—that it acts as it is acted upon is the very proof of its absolute inaction. But surely absolute inaction must distinguish that which is not from that which is: phenomenon from fact. Nature cannot at once BE, and be absolutely inactive.

R. Of course nature acts, in some sense; no one will dispute that. But may not nature act physically, and so *be* physically, and yet be inert in the other sense of not originating action? The earth *e.g.* acts in the sense of attracting.

W. Being and acting cannot be dissevered, even in seeming. To be physical and to act physically are the same; but the being physical is itself being inert, or not acting. You have here noticed a result of our endeavour to maintain *inert existence.* We are compelled at once to assert action, and to deny it. The phenomenon must act phenomenally, or appear to act, or else it could not appear to be. So a phenomenon which is felt as reality must impress us as if acting, and yet, when it is examined, be found to be inert. Thus it is discovered not to be the reality we feel it. Our perception of this passive action demonstrates an existence not passive. It is curious to see how a little difference in words enables us to overlook an inconsistency in thought. We could not say the earth acts, and yet does not act; but we can say, the earth is inert and yet acts. Is not this a penalty we pay for the advantage of using words of diverse origin? Physical things are felt by us as acting, but do not truly act.

R. Do not you forget that the inertness of matter is denied by many eminent men: M. Comte, for example, and Mr. J. S. Mill?

W. By no means. I agree with them entirely that the existence which acts upon us is not inert. I claim them, indeed, as authorities for that position. But I find their statement incomplete. They do not sufficiently observe that this existence, which they truly say is not inert, is felt as inert by us. That fact they do not account for. It is the turning point of the whole. Why is the not-inert inert to us? Is not the question as simple as this: Why is the not-moving moving to us? or, why is the earth, which is not at rest, at rest to us? Further, I think it is a mistake to speak of that which is thus affirmed to be not-inert under the name of matter. It confuses language. Matter means, if anything, surely that which *is* inert; the phenomenon or that which we feel to be. Surely those writers would not assert that a *phenomenon* ACTS; still less that the phenomenon is not material. In brief, I find nothing so simple as that nature, though it cannot *be* inert, should be inert to us, because the very essence of it is unknown. A phenomenon is inert of course. It is the same thing to say that man knows only phenomena, and to say that he introduces inertness into nature.

R. When put in this abstract form, the argument seems more powerful in words than in fact. It must be granted that inertness is inaction, that inaction is a negation, that a negation cannot exist. Also when you point out that the negation which we feel in nature is absolute, I must admit that it cannot truly be as we feel it; for absolute inaction is absolute not-being. But all this rather makes out a case for inquiry than establishes anything. There is more interest in the moral argument, for I perceive it is a question of practical life, and not of speculation.

W. I am glad you feel it to be so. We must give the proposition its scientific, demonstrative basis, and so connect it with inertia and phenomena and such unfamiliar terms. Positions which lay claim to scientific proof (which has never been held a disadvantage) must in part be treated so. But the sooner that ground can be left the better; nor need it, indeed, be tarried on; for the very same argument, which thus appeals to the intellect, addresses itself also to the other faculties of man. How could nature possibly be what it is to us, if it were in fact so little as our science represents it? Nature cannot be dead; it was called Nature because it was felt to be living. But our science seems dead enough. Are we not filled with impatience by its incessant multiplication of dead forces? Has not almost every one some contrivance in his mind for reconciling his science to his feelings? The denial of the inertness of matter may be such. Do not others say that all physical causation is the direct act of the Creator?* Others, that physical causation is not *efficient* cause, but only connexion in reason?† Do not all these things mean that nature must be more than it is felt to be by us? For if causation be God's direct act, why is it not to us as it is? Why does God's direct act affect us as inert force? If it be truly a connexion in reason, why is it to us a connexion by a passive unreasoning necessity? What causes nature to be to us different from what it is? This I seek to know. If it be not want of life in man, what is it? Of all things surely this is one of the most important for us to know; must be one of the easiest to learn. If nature is more than it is felt to be by us, do not we introduce defect into it? And if the fact of nature's laws be God's immediate act, is not God's act spiritual? It cannot be meant that God first makes inert things, and then moves them mechanically. Nor can it be meant, that there are these mechanical processes, and yet no true causation in them. It must be, that a fact not truly mechanical is felt by us as if it were so. I find all men

---

* Dr. Carpenter. † Rev. Baden Powell.

virtually assert nature spiritual, but none reflect that it must be the defect of man that makes it otherwise to him.

R. We will leave this question of inertness, which still remains somewhat abstract, and come to matters more practical. Be our state what it may, can we be other than we are, while we are in this world? In this state, man is obliged to deal with physical things—phenomenal, if you like to call them so. Must we not take up our position as it is? Perhaps hereafter we shall be different.

W. Let me first remark that I think the question whether nature truly be inert or active; whether there be in it passive necessity, or Love—true actual Being and Life, instead of deadness—can seem abstract only by my fault. If I could speak of it worthily, the words would glow with intensest warmth, and kindle a fire in every soul. If this be the fact, is not poetry infinitely surpassed; surpassed as God surpasses man? What paltry fictions all her inventions are, compared with the unimaginable truth. Is it possible that we have thought man's fancy should add beauty to God's work, exalt His world to a more illustrious grace? Poor reparation has the poet made to nature for the life of which man has robbed her. Poor, and yet man's best, and willingly received. But no—not his best. Not his best homage to the BEING God has made is rendered by his fancy, but by his steadfast pursuit of fact. Not his best worship in fictitious reverence, but in sober learning of the truth. Science is the interpreter of nature; gazes into her eyes and reads her heart. Knowledge, not fancy, shows her living. *We* clothe nature with a life and beauty from our own souls, raising her to undeserved renown! Let the thought perish, and be no more remembered to our shame! Science henceforth joins hands with poetry; they are one: the image loses itself in the reality; the shadow fades from our regard, for our hands grasp the substance. Nature is living; holy; has the life to which man shall be raised. The finger pressed no more on her mute lips, once mute, but vocal now with heaven's own music; the secret uttered, the sole secret, only to man unknown: that Life is holiness, that holiness is freedom, that freedom is necessity, that necessity is Love. God's secret, the secret of BEING, which not to know is death. Wise poet-heart, to strive, though with what vain endeavour and pitiful shortcoming, to maintain a life in nature, a sympathy, a love, a voice to human souls. The tardier science doth approve thee true, and crown thee king.

But did you ever, with great toil and patience, using both hands and all your ingenuity, straighten out a bent spring? And when you had at length achieved the task, and seemed to see your labour finished, did the unsubdued twist renew itself, and the whole thing spring back exactly as it was at first? Just so do you serve me. You invite me to begin my whole task again; all that I have said is the answer to your question.

R. Yet I must repeat it. While we are in this world, must not our business be with physical things? Must we not accept our life as God has ordained it?

W. Assuredly we must. That is what I wish above all to do: to learn what our life is, that as God has ordained it, and not under a false notion of my own, I may accept it. Let me remind you what I say. The world is more than it is felt to be by us: it is the spiritual world. This alone we truly have to do with, for it alone truly exists. We feel it as we do, because of man's defectiveness. To think of it aright, we must (as we most easily can, most naturally do), think of it as being more than merely correspondent to our impressions; remembering that we—not are to be—but are in the eternal world, and that the fact with which we have to do is the raising man to his true Life. Perhaps you decline to adopt this view; but at least be just. Do not say it is difficult, or unlike things that we commonly admit. What is more constantly our habit than to remember that our apprehensions are inadequate, and that the truth of things differs from our impressions? Do we not always add in our thoughts to that which, strictly speaking, we perceive? When we look at a chair, for example, do we suppose it to correspond to the impression on our sense? Clearly not; we see only parts of it: *such* a chair could not be; we infer, and conceive as existing, that which is not to our sight. We supply something wanting, the unseen portions of the chair, and then the thing is possible. What more easy than to do the same by nature, to add to it in our thought something that is wanting in our impression; to remember that our perception implies an existence unperceived, that the true existence is active, not unacting: living and not dead?

R. I cannot say that is difficult.

W. Then, if you do that, you distinguish between that which truly acts on you, and that which seems to do so. You understand that your feeling yourself to be in an inert world is a false feeling on your part.

R. That is, I consider that which *exists* to

be very unlike that of which I have the impression.

W. And the cause of this difference of your impression from the truth is a condition of your own, which you share with all men. In considering the relation of the whole human race to the existence which is present to them, you have regard to the condition of the whole human race. You reflect that man is defective—that all men are so—and that all therefore perceive the world as defective.

R. This may be very well in theory, but the practical difficulty remains:—that the world we feel and know does not exist. How can we believe it?

W. We feel these material things: we know them to exist? We are quite sure of it?

R. Yes.

W. An ignorant man feels that he knows many things; he is quite sure they are true: that the sun moves, or that smoke rises because it is light?

R. True: the more ignorant the man, the more sure he feels.

W. An ignorant man feels himself to know, but the things he knows are not true. Is that knowledge?

R. No. That is ignorance.

W. Suppose that we feel certain things to exist, but the things we feel do not exist: what would that be?

R. It would be, in respect to our life and being, as ignorance is in respect to knowledge: an absence of it. I perceive the argument, but you have to make good that the things we feel to exist do not exist.

W. That cannot be difficult. How can a world

answering to our inadequate feeling and apprehension exist? That which has not action is thereby proved not to be itself an existence, but an appearance produced by the action on us of something else. A world such as we feel can no more truly BE, than a chair could stand erect with the two legs we might happen to see: and for the same reason, viz., that our apprehension is inadequate. Why should it more influence our thoughts that we feel a world that is inert, than that we should see a chair with only two legs? If that which answers to our impression cannot exist, then our impression does not correspond with the truth. There is no maxim more thoroughly familiar in practice, or incorporated into our habitual thoughts, than this. Therefore I say that we can now perfectly well know the world to be spiritual, and deal with it so: knowing that our apprehension of it is defective and why it is so. That is only to recognise our true position, not to alter it. Let me take another instance: suppose it had been said, in reply to the assertion of the earth's motion, that while we are as we are the sun must be revolving *to us*, and that we should continue to think and act as if it were truly so: were not that ridiculous in theory, absurd and hurtful in practice? But not more than that we should regard the world as truly physical, because while we are as we are it must be physical to us. I say that we may know that it is not truly so, and that our thoughts and practice must conform to our true knowledge, not to our false impression. As astronomers treat the earth as moving, though they feel it steadfast, so may we treat the world as spiritual though we feel it inert. Is not thus acting according to truth, and not to appearance, always what is meant by common sense? Is it not opposed to reason to say, first, these things are but phenomena, and then to say, treat them as the realities? That which is but a phenomenon, or appearance produced by something else, I will treat so; I will treat it as it is. But there is truth in your remark that while man is as he is we must feel as we do, and have the experience we have of a physical, material life. And hereafter, assuredly, we hope to be different. But let us not mistake. The question is, how we should act *now:* what is the truth of this present state? The answer I give is: remember man's defect of being, which alone makes us feel as we do, making phenomena realities; and treat the world as it truly is, as spiritual: have regard in all things to man's redemption. When man is made alive then phenomena shall be no more realities to him. The alteration from physical to spiritual must be by a change in man. Our being in this physical state, which we would make an excuse for treating things as we may know them not to be, is the result only of his defect of being. We shall be different when that is done away. Can we not easily understand that as, by defect of knowledge, man has opinions not corresponding to the truth; so by defect of being he has feelings or perceptions not corresponding to the truth? For example, a revolving sun is the fact to his thought till he knows more; an inert material world is the fact to his experience till he is more.

R. This is the very point. There is a difference between the cases. An opinion is a thing which we can alter; we understand how we are and must be under mistake about things; but our experience and feelings we cannot change. The material world is real to us, as you say; then should we not treat it as being real to us? What other reality can we have anything to do with? And again with reference to your illustration of the chair of which only two legs are seen, we have a means of knowing how it truly is. We can see all four legs by looking, or we can feel them. There is not a parallel. We cannot feel the world not to be physical as we can see a chair not to be of such a form as it presents to our sight. We always feel the world inert.

W. These are the very things that will make clear my meaning, and show you that we are not really opposed. All that I seek is to alter an *opinion*. We have thought that the reason we perceive and feel as we do is that there truly exists such a world as answers to our feeling. I say let us think that the truly existing world is more than it is to us and that we are defective. This cannot be impossible. I seek only to understand our experience, not to alter it. If it were altered, what I say would be no longer true. The whole position of man's want of life is founded on the fact of his experience being of a physical, material world: in a word, of a dead world. Suppose the sun were not any more perceived as moving, would not the proof of the earth's motion be destroyed? So would the proof of man's deadness be, if nature were not felt by him as inert.

You say that we have means of correcting our impressions respecting individual things but not respecting nature as a whole: there is here however exactly the difference there ought to be. We cannot correct our impression of the world as inert, by means of our senses: it is inert to all our senses unitedly, and in every use of them; but we can correct it by means of our thought. We have an intellectual as well as a bodily perception. The laws of thought, equally with the feelings of the sense, determine our opinions. We can as certainly know that our impression of the world as inert differs from the truth, as we can that any other impression does so. We must use the appropriate means: due consideration and a right use of reason. We must reflect whether it is possible that our feeling should not be defective; whether that which is but a phenomenon and not the very fact that exists must not necessarily be unacting; whether it is not absurd to infer that therefore the very fact must be unacting also. In short, we must consider whether the argument, that the world is physical to man only by his defect, be not good and sound. There is no other way in which the question ought to be decided. Nor is the question in itself peculiar. It differs from other cases of the use of reason only in its bearing and results, not in its nature. For by what means do we distinguish the truth from our impressions in cases in which we cannot apply the senses: in astronomy for instance? Only by such use of argument and reason. But this is not the end. When you say that we cannot *feel* the world not to be inert as it is to our feeling, I join issue with you entirely. I affirm that we do most emphatically feel nature not to be that dead, inert mechanism which it is to our conceptions. I appeal to all the history of human thought, to literature, poetry, science; all are leavened with no intenser feeling than this which you deny: that nature is more than we feel it to be. I would appeal to science above all; for all its history is a strife between these two feelings: that nature is living, and that it is dead. All the strangeness and repulsiveness with which it affects uninitiated men, all the strife which it undergoes in extending its domain, the ever-renewed collision between it and the devout affections, are due to this two-foldness of our feeling. It cannot be mere DEAD necessity that constitutes this wondrous life. It is no matter that we feel it so; we feel as much that it is not so. Oh happy reconciliation of a strife too long and weary; peaceful end of a contentious toil; bright recompense of zeal undaunted and ungrudging labour; that the deadness is within, the life without. True it is, we FEEL a deadness and we FEEL a life. What shall we say? How shall we apportion them? Is the deadness man's, man's only? May we, dare we, think so? Is this the consummation of the hope, the resolution of the doubt, the interpretation of the mystery? Man wrestling so long with nature, to gain this victory: to know himself? 'Tis rest, and energy; 'tis humbleness, and exaltation; 'tis content, and hope unbounded; 'tis self-renunciation, and high resolve; 'tis penitence and joy. Let me bow my head in shame, it is delight to be abased; let me lift up my soul in joy, I will exult in God. Thou narrow and contracted heart, seeking thy own good, labouring fearful and in doubt, expand thyself, cast off thy shackles, melt and be utterly dissolved away. This is death, not life. Let glad laughter take the place of tears, and energy, new born of joy, chase weariness for ever. Oh sacred Life, that bearest us in thy bosom, swelling around our empty souls that shall be filled with thee; in thee we do rejoice. Man's life, his hope, his destiny rise so much higher to our thought. Because our aspirations were not large enough, because we were too easily content, because we mistrusted God so much and hoped so little; therefore the world has been so dark. Our LIFE is more than we have dared to think.

R. I cannot blame your enthusiasm. If I shared your belief I should also share your joy. But the question now is, not what is beautiful, but what is true. In speaking in this way about life and death, are you not confounding words, and introducing perplexity instead of giving definite knowledge? We are living now, and we die when the breath leaves the body. These words may of course be applied, figuratively, to other conditions, but you do not seem to use them so. It is difficult to make the thought follow you.

W. When was a new conception, however true or simple, first introduced without such difficulty? How long it takes a person to whom the idea is strange, to understand that at the antipodes people have not their feet uppermost. And the feeling has every justification. Nothing can be plainer than that it is entirely a new use of words to say that they are not head downwards; in no possible case can the evidence of sense be more complete. Yet the whole secret lies in this: that up and down are relative terms. So are life and death. So indeed are almost all our words. It introduces no perplexity whatever to think of this physical life and death as having its relative place within that true deadness which constitutes man physical. All our thoughts in so far as they are disciplined, or approach towards accuracy, are moulded into this relative form. Do we not think of things as being at once large and small in different relations, or as relatively true yet absolutely false? May not a thing be truly moving, yet relatively at rest, so that we consider it as either, according to the relations in respect to which we regard it? Even so may man be truly dead, yet relatively living, and be considered as either, according as he is regarded in relation to the absolute, or to the phenomenal—to the true life or to the apparent. I deny that there is any perplexity here, or any laxity in the use of language. I use the words life and death because I mean the things. And for justification I appeal to every literature. What tongue is there in which a life and death of man, apart from bodily life and death, is not recognised; another relation of man than to the physical?

R. But you seem to invert the natural order of ideas. We have been accustomed to regard the life which we are conscious of as being primary, and as the basis of all:—that we are truly living as men, but, according to the state of our feelings or our will, we have or have not a life to the spiritual. You seem to imply something the opposite to this.

W. I admit that I seek to associate a new thought with the words life and death. But am I not right in doing so? Do I not rather restore, than invert, the truly natural use? Have I not made it good that this physical life, with consciousness of the self within and perception of external deadness, is not the true life of man: that it arises, and must arise, from want of life in him? If this be true, then I am right in speaking as I do of life and death. If it be not, then my whole thought is wrong, and my use of these words is only part of a larger error. Words must follow thoughts, although they may also lead them.

I say that if we perceived things as they truly are, we should consciously perceive that man is wanting in life; even as, if we were removed from this earth, we should perceive that man is moving. And that we may now think and act according to the truth, unembarrassed by our false consciousness. Our consciousness of life, when man is not living, need no more perplex us, than our consciousness of rest when man is not at rest.

R. You would have us take a view apart from our own mode of feeling, and rise above that which is perceived by man as he now is, to that which would be perceived by him if he were different. And you appeal to astronomy as proof that we can do so. We do so by recognising something in our own condition which affects the way in which we perceive.

W. Exactly. All our perplexity comes from making our consciousness the measure of the reality, instead of recognising it as the measure only of the phenomenon, from which the reality is to be learnt by well-directed investigation. Assuming that which is consciously present to our perception to be that which IS, we can comprehend nothing. Intellectually the world is a mystery to us; morally a fearful problem. Am I wrong in saying that there is a remedy for this state of things in remembering that man's deficiency modifies his impressions and necessary convictions, and in endeavouring to ascertain in what respects the very fact that IS must exceed what is to his consciousness. That is, technically, to find out and exclude the negative element in his perception by taking it into himself. If we will do that the clouds roll back, the mists clear off, the darkness turns to light. Nothing is altered: but we understand. Our daily work remains the same as before, but it is done with a new spirit. We direct our aim to the reality, using the phenomena with reference to that, treating them not falsely, as for or by themselves, but truly as they are, in relation to a different fact which alone causes them to be perceived. We live for man's redemption. We see that the raising man to true and worthy life is the secret of human experience; the sacred mystery of nature.

R. Here let me ask you another question. There is a want of coherence in your language. You say, first, the cause of our experience is our presence in a defective state in the spiritual world, and then that it is a raising man to life. Do you mean to say that these are the same? If I granted that a spiritual object (as defined by you, *i.e.* a not-inert one) is the cause of my perception of an inert object, say a table *e.g.*, how can the redemption of man be also that cause? When I ask of you: what causes me to perceive a table? I might receive as a fair answer, 'the action on you of an object different from a table, and especially in this, that it is not inert as the table is;' but what can I think of such an answer as this: 'Man's being made alive causes you to see a table:' and especially what am I to think when both these answers are given at once? To me they seem, to say the least, by no means the same.

W. I owe you more explanation on this point, but I think you will easily see that the inconsistency is only apparent. Let me revert to the ordinary idea, that our perception is caused by the operation on us of physical things. Now I ask the same question: What causes me to perceive a table? and you answer me, 'the action on you of the table.' But again I ask you, this time as a physiologist: What causes me to perceive a table?

and you say, 'a certain condition of your nervous system, some molecular operation in your nerves and brain.' Ought I to charge you with inconsistency? These are two views of the same thing. My two positions, that the cause of our experience is the action on us of spiritual existence, and that it is the raising man to a truly living state, are two views of the same thing likewise.

R. But you do not make clear your view with respect to our perception of physical things.

W. My expressions may have seemed obscure, because they were meant to be general. I do not give any opinion as to the details of our perception, nor do I attempt to separate perception of physical things from the total of our consciousness. Of that consciousness as a whole I say that it must be due to the action on us of an existence not inert (or physical), and that our having conscious perception of inert things as the reality demonstrates a defect in man. But I do not go farther. Possibly we may hereafter attain sufficient knowledge to enable us to understand why our perception must be in all respects such as it is. But that is a question which must be solved by sedulous examination and study. We should ill have profited by the past if, directly a new problem is presented to us, we began to guess.

R. I understand you to say that you only affirm, in general, that the Fact which truly exists is spiritual, and that the action of this true existence upon us may be represented by saying that it is the making man alive. This is man's relation to the spiritual: but you do not attempt to go any further.

W. Yes. I do not entertain any opinion why the truly existing (or absolute) should cause us to perceive stars and planets, or earth and water, or trees and animals. I think some dim intimations of why it should be so may be gathered, some guides to investigation feebly grasped, but all such questions clearly must remain. They do not press for solution. They do not bear upon the question whether the perceived defect is in man or apart from him.

If I might illustrate my meaning again, I would say that the proof that the earth revolves, and not the heavens, is entirely independent of any question about the nature of the starry universe, or the reasons of the planetary courses. These were problems for future investigation, and even yet they are but begun. But the knowledge of the earth's motion was the indispensable basis for the commencement of the researches which promise us, in these respects, so ample a reward. So I conceive that the recognition of deadness as man's, and not as nature's, is a basis indispensable for the commencement of an investigation as to what nature truly is, and why we must feel it as we do.

R. When you say that the spiritual world is the Fact which causes me to perceive a physical one, I must neither suppose you to mean that chairs and tables are spiritual, nor that there are spiritual chairs and tables, of which these are the images, as perhaps some Neo-Platonists meant; but simply that my perception of these phenomena is due to the existence and action of Being that is different from them, and of which we can know that it is certainly spiritual—that the inertness we perceive cannot belong to it,—but respecting which you do not pretend to say why it should cause us to perceive as we do. We must, in a word, leave the particular relation of the phenomenal to the absolute to be investigated, if it be found capable of investigation.

W. Precisely so: the effect which the absolute must produce upon our consciousness involves the three elements—1st of what *it* is, 2nd what *we* are, 3rd the relation between us and it. If we knew more of ourselves and our relations to the absolute fact of nature, the effect on us, of which we are conscious, would surely enable us to know something respecting it. But we must not hasten. Nothing is truly so unreasonable as our habit of inferring causes directly from the phenomena we perceive, in cases in which our knowledge is not complete. I was struck with a trivial illustration of this a short time ago, when a bright circle of light suddenly darted around the walls of my room. The children were delighted. I do not know what they thought of the cause, but I should never have imagined it myself, without a good deal of experience of the deceptiveness of sense: a man was carrying a tin can past the window in the sun. Now it is easy to see how this cause should produce such an impression on me when I understand the circumstances: the sunshine, the reflection of light, and so on. But suppose me without such knowledge; how then should I possibly have inferred a tin can from what I saw? I cannot even now detect the slightest resemblance between what was present to my consciousness, and that which caused it to be so. I could not help thinking what an infatuation possesses us, when we fancy we can immediately infer from our impressions of nature, from that which is present to our consciousness, what that is which causes it to be so. I wondered that I ever should have inferred, from the aggregate of my impressions, that the cause of them is a material world corresponding thereto. I am by far too ignorant. This only I venture to say: that if we could ascertain all the circumstances, we should see that our impressions ought to be such as they are, and should be able to trace how they must arise. Even as I can, in scientific fashion, trace my impression of that meteoric flash to the sun and the tin can.

But I bethink me of another use of my illustration. Have you not watched children, sometimes, trying to catch such flashes? Alas, my friend, that is not only done in play and amid merry shouts of laughter. My ears are filled with groans and blasphemy instead, and faces pale with care, and scarred with passion, rise before my eyes. The scene is changed; but not the actors nor the game. Game do I call it? it is grown to deadly earnest; a mad battle for the glancing shadow. Hope and despair; triumph and rage; hatred and envy—Let the scene be closed. These are our Brothers that we look upon—ourselves. Will no voice warn us? Shall we never know? Never, like grown men, turn from the Appearance to the CAUSE?

R. I understand you then to say, that the Fact which causes all these things to be present to our consciousness (*i.e.* the absolute) is spiritual. That in relation to us this spiritual fact is the making Man alive. Consequently that to regard things as they truly are, and to act according to the reality and not to the mere appearance, we must in all things consider and have respect to the redemption of man. That is the reason which necessitates, and is the only true cause of, all our experience.

And you say that our natural, and as it were intuitive, conviction of the true existence of these inert things is due to our natural ignorance: being just such a conviction as a person looking through a stereoscope, without knowledge of the circumstances, would have of the existence of a solid body such as he has consciousness of. And as to act aright, or to succeed in his action, in reference to the object of his vision, the gazer through a stereoscope must act not according to his impressions, but according to a knowledge founded on examination and reflection; must act with reference to something different from that which 'is to him'; even so must we. If we would practically succeed, we must treat that with which we have to do as being of a different kind from that which we are immediately conscious of. We must guide ourselves by a knowledge which subordinates our natural impressions.

W. Even so. If we separate the phenomena, the things that are present to our consciousness, from their connexion with that absolute existence (of whatever kind it be) which is truly acting upon us to make us perceive, and then ask whether those phenomena exist, of course they do not. It is like asking whether a single solid thing exists in a stereoscope, ignoring the pictures; they both do exist and do not. They exist as phenomena, as that which we are made to perceive by the truly existing things, but can only be regarded aright when viewed in that dependence upon, and relation to, something else. The entire difficulty about the material world arises from this unnatural disconnexion of the phenomenal from the absolute. Let them be rejoined, and nothing can be simpler. The physical world exists as a thing that we are made to perceive by our relation to the spiritual. It has this existence and no other. There would never have been any need to discuss the existence of the phenomenal, if such a false, isolated existence, apart from that which is not phenomenal, had not been asserted for it, through our ignorance. Any person may see the nature of the case directly, who will suppose himself, through ignorance, convinced of the physical existence of a solid body in a stereoscope, and another person denying it, and trying to make him understand that there are two pictures instead. Let him conceive that the denial of the solid body seems to him like a denial of common sense, like affirming that there is nothing there at all, but that all is an illusion, and that to refer his impressions to two pictures and the laws of his own vision seems to him absurd: then he will perfectly realize the nature of the difficulty which is felt when the physical world is denied to exist, and our consciousness of it referred to a spiritual existence and the state of man's own being.

R. It is only through ignorance that we are so convinced of the existence of physical things? That is a natural impression which needs to be corrected by learning the true circumstances of the case?

W. Yes. It is not they that exist, but something more and better than they.

R. But it is difficult to avoid being confused by these illustrations. The difference is so great between one material thing making us seem to perceive a different one, and that which is not material making us truly perceive things which are material.

W. Hold fast to the difference. If the cases were the same, how could one serve to illustrate the other? The one relates to merely phenomenal conditions, and to the order which the intellect demands in its conceptions, the other to the very being of man. The one relates to portions of our relative experience, the other to man's experience as a whole. Yet these instances of known sensuous deception are true to the point in hand. They help us to understand what the nature of perception is; to recognise that what is consciously present to our perception must depend on what we are, and how we are circumstanced. They should at least suffice to break through the only real obstacle to an understanding that the world is not physical, viz., our firm persuasion that what we set down as our consciousness cannot land us in a false conviction.

R. There is more besides. This physical world is so very unlike what we hold the spiritual to be. Ask yourself like a reasonable man: How can our being in a spiritual world make us live upon a solid earth, build material houses, eat material food, do all these unquestionably material things? Why should we not think that we are spirits encased in material bodies, as we have always believed? It makes everything so simple. That I——I was going to say—*that* I understand.

W. It is well you stopped. Do not you see that this is exactly the hypothesis, or supposition according to the appearance, of which it is the nature to seem simple at first, and be found mysterious and incomprehensible at last? That is to the intellect the broad path that leads where it should refrain from going. We must be content to enter the strait gate. In respect to knowledge, as to life, heaven is inexorable: the path is only one. We must submit to use reflection and thought, to be guided by evidence, to incur trouble, to set aside convictions however cherished, if they will not stand. We must begin so. The entrance is hard; not the end. The end is liberty, and light and gladness. If I could make you feel what it is to know that man is wanting in his life and that we *are* deceived, you would not argue with me any more.

R. But it is hard to conceive that these solid things do not exist. We take them as the type of existence rather; and when we say of any other things that they exist, we mean that they are as real as these.

W. It is not exactly so. Of some things we are obliged to say that they are more real than these. And in truth the difficulty is not so great as it appears. It is the substantialness of the world that makes it real to us; that we work and walk about in it. In fact, it is its *existence in space*. Did it ever occur to you to ask yourself what space is? or how man arrives at the notion of it? Reflect for a moment. Is not space exactly negation—absence of existence—pure and entire 'not-being'? We cannot think of utter absence of being under any other mode than that of space; we cannot exclude space from our thought of absence of existence. For when we try to exclude space from our conception, we have to think of BEING that is not in space; as spirit is held to be by some. Is it not a striking thing, that we have obtained from our experience of the physical world an idea which, when we examine it, we find to be that merely of not-being; that this not-being is the very essential condition of the physical, under which only it can exist? Let us not scruple to use our reason. Surely the feeling and conception of *space* is the very one which we ought to receive in feeling that to exist which does not exist. Space, or not-being, may well be the condition, or mode of existence, of the 'phenomenon.' It is like the inertia, or not-action, which we also associate with it, and tells truly the same tale.

R. It is of course the occupancy of space that makes things solid, makes them to us realities.

W. It is their 'existence in not-being!' that which we feel as real demands 'not-being' as its condition! Do we not find out by this curious linking of our ideas, that we are feeling that to be which is not? Or again: let it be supposed that man feels that to exist which does not truly exist: can we think of any other way in which this could be, than by means of that very solidity which we associate with existence in space? So inert things are our realities. Their existence is not truly existence, and we aptly term it existence in space: it is the reality to us of the unreal. Man feels himself in space by his defectiveness; by his want of being, 'not-being' is felt around him.

R. Let us leave these abstract thoughts. I grant that I cannot prove the existence of any world at all by the present metaphysics.

W. Analogies help us more, and they are never wanting to anything that is true, for nature lends all her treasures to adorn whatever she acknowledges. What we feel so strange is that we should perceive around us so definite and substantial a habitation as this earth, if the physical does not exist absolutely, but is merely the phenomenon to us of some other existence. But look at the sky at night. Consider the firmament. Is it not stretched as a canopy folding in the earth, of definite circumference, and solid look? Do not say no; for humanity would testify against you.

History proves that it appears so to man's natural eye. Is there any such canopy around the earth? Is there anything like it? Man dwells, to his consciousness, in an encircling heaven which is not. A habitation, bright with gems and stretched on everlasting pillars, has been prepared for him; —by what? By his presence to infinity bestrewn with lavish worlds. And why? Because it is the nature of his sight. Why should not man's presence to the spiritual infinitude of being place him, to his consciousness, in a home like earth, amid a universe of stars? Do we ask why? Because it is the nature of his present state to feel as dead that which is living; because the phenomenon which he perceives is different from the truth of things, and by his defect of being the phenomenon is his reality.

DIALOGUE II.

R. I clearly see your meaning: one thing acts upon us, and another is consciously present to our perception. The former you call the Fact, and assert that it is spiritual or active; the latter is the phenomenon, and it is physical or inert. The spiritual truly exists, the physical exists only as an appearance. If man were in a truly living state—not defective in his being—he would have feelings correspondent to the truth; but inasmuch as he is defective, his feeling is wrong. He feels that which is only phenomenal to be, if not the sole, at least a true, reality. And then, thinking according to this false feeling, he finds himself in entire perplexity, and unable to understand the very being of anything. You would say that he needs to direct his thoughts according to a different plan.

W. If we directly know only phenomena, what is the practical inference? How should we learn, or try to learn, the Absolute from them?

R. Of course we must have regard to the state of man, and consider the things which falsify his impressions. I concede all you say on that general question. Nor indeed is it new. Only in the application of the principle do you differ from others. For the common idea of the world, as consisting essentially of matter and force, also involves a consideration of man, and how things are altered to his perception. It reduces, for example, sound and light to mere motions of particles, and supposes in man such a 'nature' as causes him to perceive this bright and variegated and musical and odorous world, through the action on him of something wonderfully different.

W. Quite true. I only wish to apply an established principle. If any one will take up this question simply on its merits, I cannot doubt that he will agree with me in thinking that nature must be *less* to us than it truly is, and not more. And that we have only for a time fallen into the other way of thinking of it, because of our ignorance. Surely any idea, which enables us to escape from the necessity of supposing that matter and motion mysteriously affect us with such perceptions as we have, ought to be welcomed as a great relief. We get accustomed to such views, and so lose all sense of their amazing difficulty, and quite fail to remember how they would impress us if we now heard them for the first time.

Thus the question stands: Nature is not truly and in itself such as it is to our perception. This is common ground: nature is altered to us by man's being such as he is. What then is that in man which alters it? Is it his known defectiveness, making that which he perceives defective; or is it a power in him of adding innumerable qualities? According as we answer this question will be our inference from our perceptions. If it be the latter, then we infer an inert matter and endless forces different from what we are conscious of: if the former, then we infer an acting, spiritual world different from what we are conscious of. Is not the relation between these views evident in the mere statement? Are they not, respectively, hypothesis and truth? Is not the former the first natural construction man puts upon his experience in his ignorance, before he has learned to read it aright? Does he not cling to it now, as he has clung to all other such natural errors, for no reason but that he is accustomed to it?

R. You might strengthen your argument logically by asking also whether it be, on any ground, admissible to assume many positive elements in man, as altering nature to him, when one known defect will serve the purpose. There would be no reasoning at all, there would be no more any science, if one known cause could not demand to be received, instead of several merely assumed to account for the phenomena.

W. You are right. Man's defectiveness is known. That is a point on which all schools of thought agree. So that my position is this: Nature is made to differ, to our consciousness, from that which it truly is by a known cause; namely, by man's defectiveness. That defective or inert existence therefore, which we conceive, is not the truth of nature, any more than that which is to our sense.

R. That is, indeed, only saying that science deals with phenomena alone, and that man's thought does not penetrate to the Absolute.

W. But observe: I affirm of the Absolute not only that it is not inert or material, but that it is spiritual. Here I fall back upon the conscience in man. I say: If we admit that the true reality of these phenomenal processes is not an inert necessity, but is true action, then we must admit it to be *right* action. Science proves the true being of nature holy, in proving the phenomenon of it inert. If there were not necessity, or holiness, in the Fact, how should there be necessity, or inertia, in the appearance?

Is it possible for one who admits undeviating Law in the phenomenon, that is, who admits science at all, to deny Holiness in the Absolute, unless he either affirm that Absolute to be inert, or else refuse to let his words express the unavoidable workings of his consciousness? Can we separate moral quality from action truly so called? Observe, I am not asking now for a verdict respecting what the Absolute IS, but what we must think according to the laws of our consciousness. If the action in the absolute were not always one, would not the passive change in the phenomenon be variable, and science be impossible? 'Inertia' is but the phenomenal reflex of holiness. If the acting Absolute in nature were not holy, man could never have constructed a science of the unacting phenomenon. Is not that Holiness, indeed, the true ground of our confidence in the universal prevalence of Law in nature, which has never yet received any adequate explanation?

R. I do not call this in question; granting that there is a spiritual world at all, of which we will speak hereafter. If you can prove that the cause of our consciousness is not a passive, but an active existence, few will refuse to follow your inference, that the invariableness perceived in the phenomenon implies, according to the necessary conceptions of mankind, a moral necessity in the action. Prove the spirituality, or activeness, and the holiness will not be denied. I do not, myself, know of any class of men who wish to ignore moral distinctions, or who would attach less value to the necessary conclusions of the moral, than of the intellectual, sense. I think those against whom this is charged are for the most part misunderstood. The question rather is: Does the excluding the perceived inertness from the Absolute truly involve its spirituality, in the sense of such activeness as we can denote holy?

W. I should be most happy for this conclusion to be tested in every way, but I cannot myself think at all in any other. I cannot even conceive any alternative, or possible third course. By saying that there is true action, I mean holy action, and cannot suppose myself meaning anything else. If the moral element be rejected, I am landed in inertness again at once. On this point, therefore, I am wholly in your hands. But if the argument needed re-inforcing, might we not appeal to that which nature is, to the wonderful processes and results of the organic and inorganic laws, and ask: If there be not passive necessity as cause of this, then what cause? if neither mechanical, nor spiritual, necessity is here, then what necessity at all? how can these things be? I should say: Inert necessity *appears* to account for the course of nature: Holiness does truly account for it. But besides these, what account can possibly be conjectured? Especially, what account which does not re-introduce the banished inactivity? If it be replied that we are not called upon to account for it at all, I should make answer that all systems do give account of it in some mode or other, and that if my argument be allowed, we have agreed that that which exists is not inert. We have gone so far as to account for the course of nature by true and not by apparent, or passive, action.

R. I should not dispute this point if I were satisfied of the rest. But I remark a difficulty you labour under in respect to words. You do not mean to affirm of the Absolute all that we associate with the word 'moral.' Not such moral action as ours; such holiness, maintained against temptation and in spite of self; but a holiness from which these elements of strife are banished; a true, spontaneous, necessary holiness, such as we hope for in heaven, such as we adore in God. You are obliged to use the word moral, but the idea it conveys needs elevating.

W. The word spiritual is better. That seems to me to express the true conception. I might define it as that to which holiness is necessary. Man's toil and struggle to be holy arise from want of the spiritual in him; they arise from self. There is true holiness in nature because self is not: there is no liking evil, which alone makes 'virtue' possible. But in one word, Nature's necessity is Love. Holiness is action made necessary by love.

R. But if the inertness we perceive be not truly in nature, if that particular defectiveness be due to man's condition, still it does not follow that there is not some defect in nature. We do not thereby assert its absolute perfectness. I presume you would admit this.

W. Certainly. All possible questions of that sort remain open. I affirm only that the phenomenon alone is inert or material: the true existence of nature acts, or is spiritual.

R. I see that our inference of a material world rests upon our perception of inertness. Matter and motion are what we must infer, assuming our impression of nature as inert to be correct. And the question you would have us ask is: Why are our impressions such as they are; such that, not recognising them to be influenced by man's defectiveness, we have necessarily inferred the material world, with all its properties and forces?

But do we not here come to this fact: that we

are ourselves conscious of *moving?* This consciousness of motion is the chief ground of belief in matter and motion as constituting the world.

W. True. That which we feel to exist is in space; that which we are conscious of involves motion. I do not deny the *materialness of the phenomenon*, or of that which is to our consciousness. Keep your eye steadily to the point. Nature is material *to us;* we consciously move in it, and must do so. But the question is: What is the true cause of this consciousness? What is nature apart from us?

R. You mean that there is not truly motion, although we are conscious of it: that which truly exists, and makes us feel motion, is different. You treat motion as we have treated luminousness, and say that we are made to be conscious in that way by something that does not resemble that which we are conscious of.

W. That is what I mean. I have, at least, examples to urge in support of it. Motion is a mode in which we feel something that is not itself like motion. But again: evidently the question of *motion* and that of *space* are one. As space is a condition of all phenomenal existence, so does motion seem to be of all phenomenal action. It is curious that we cannot think of any natural action, except under the form of motion. Some men argue, absolutely, that all material processes can be nothing else than forms of motion, and for my own part I profess an inability to *conceive* them in any other way than as motions, either of larger or of smaller particles. This fact is full of instruction. Motion, in this respect, agrees with all those qualities which are introduced into our perceptions by our own condition (all subjective qualities), viz. in being universal. It applies to all our perceptions: as any condition must which belongs to ourselves. I would suggest that it is a condition of our own that necessitates our conceiving all perceived actions as motion. Motion is that which all action becomes to our conception. All the action of nature, of whatever kind, is motion to man's thought. Light, or sound, or warmth, everything which he perceives, refuses, when he endeavours to conceive it, to be anything but motion to him. And indeed, if we consider, it is evident that the mere fact of man's consciousness of space necessitates this. All that he perceives he must refer in thought to action in space, or in respect to space, which is motion.

R. Perhaps there is something in this. I do not expect you to be able to give an explanation of all things, and the question of motion may want further examination. I see that it is only part of the general question, and does not specially affect the inquiry whether the Absolute be spiritual or not. And I do not wish to be one of those men, of whom William Harvey says, that 'they will not receive a new system unless it explains everything.' It is surprising how natural it is to adduce any unexplained circumstance as an argument against a new view, without considering whether that view ought to explain it, or whether it is better explained the other way. All of us have a feeling, as if an opinion we have before entertained ought to be held, not only until there is sufficient evidence in favour of another, but until that other has given an explanation of every question we can ask. I have learned to be on my guard against that weakness.

W. Harvey could not explain why the arteries were found empty after death. At least he could only suggest probable reasons. The idea that they contained vital spirits accounted perfectly for that circumstance. Yet it was proved that the blood circulates, although the reason of the emptiness of the arteries remained to be investigated.

R. It may be proved, you would say, that the Fact of nature is spiritual, although many things cannot be accounted for.

W. Observe: I do not deny that 'we move;' but I inquire what is the true meaning of that statement. There is motion, of course, in the same sense as there are physical things. Motion is either truly existing, or is phenomenal only, according as that which EXISTS is truly physical, or only phenomenally so. The question is whether our consciousness is to be assumed as correct, or whether it is to be investigated and accounted for. And I would suggest farther, whether the right question for us to ask respecting all our experience be not this: why *we* are so impressed; why *we* have been obliged to make such suppositions and entertain such convictions? In connexion with the material world, the sole fact is that men have been, and we are, necessarily impressed with a conviction of its existence. But why should it be called a 'melancholy' fact if there be not such a world? What should follow but that there is a better one? Where is the harm if we are naturally under illusion? When once we know ourselves to be so the practical evil is at an end. Or if we erect our impressions into an authority, and say, 'There is a material world,' what are we advantaged? We have gained for our belief a world of low inferior order, one that even we cannot but feel to have some evilness and degradation in it. If we will admit a different plan of thinking, and consider man's known defect, then we may believe a world at least excelling that in value, one that is at least perfect to *our* thought. It is every way a gain. Is the sun less bright, the earth less solid, food less satisfying, are smiles less sweet, or words less full of meaning, to one who believes that the world is different from that which it is felt by him to be, than to another? Is his confidence less in the stability of the natural laws, because he refers them to an absolute holiness, instead of to mysterious 'properties,' of which he cannot know that it may not be the property to alter, or at least to produce different effects, to-morrow? It can, at least, be no loss. Our sensations are not altered by the change in our views, as all agree in urging. What then is the difference? the difference to thought and belief, in respect to which alone a difference exists? Is it not wholly an advantage? One has to his belief a low dead world, not to be understood, with some strange badness in it: the other a world infinitely glorious, thrilling his soul to ecstasy, and a conviction of deadness in himself that rises into aspiration towards a worthier life for all.

For is it not evident that we need not affirm defect without us, if we will admit it in ourselves, in man? Suppose defect within: will it not be perceived without? And to perceive defect without, what could that be but to be in a material world, or something essentially the same? I will not go so far as to say that a Being who, by defect within, perceives defect without must feel himself surrounded by, and embodied in, matter; but I can think of no other way in which there could be perception of defect as external. Must not such a being feel inertness in his world, and be conscious of exertion, and of force? Would he not necessarily infer matter, and suppose 'inertia,' and construct a science of passive laws, based on the fact that the action around him did not vary? He would think nature dead, nor ever ask himself the question whether it could be truly so, until he had exhausted all contrivances to maintain his natural impression as the truth.

R. As to our being conscious of moving, it occurs to me that there is what you call a phenomenal illustration, which might help us to understand it. We feel ourselves conscious of being steadfast, and of the earth being at rest, yet we have good reason for believing, not only that it is not so, but that there is no such thing as that kind of rest in the universe, and that all the stars are in motion. If we can be conscious of steadfastness while there is truly no such thing, why may we not be conscious of moving though there is truly no such thing as motion? What we are conscious of and what truly exists are different questions.

W. I thank you for your suggestion. Parallels of this kind are at least of avail to meet objections that are apparent only; and help to clear the course of an argument from considerations which do not truly affect the conclusion. And let me point out to you how your remark also bears upon what you just now observed, that Nature may not be wholly free from defect, although not having in it the defect which we perceive as inertness—although that may be from man. The sun and stars are moving, although the obvious motion which man at first attributes to them is due to his own. We need first to understand their relative steadfastness, and our own motion, before we can begin to examine whether they truly move. So, perhaps, we may inquire respecting defect in nature, when we have first recognised its relative perfectness, and man's defect.

And while this idea is before us, let me remark an application of it to the question of freewill. We feel conscious of an arbitrary freedom. Yet perhaps it may be, not only that man's arbitrariness is not freedom, but that there is and can be no such freedom; that man feels himself conscious of a thing that cannot be, and that in fact freedom excludes arbitrariness.

R. That may be true: at least it is beautiful to think. Then God, as of all Beings the most free, is also the most free from arbitrariness. With Him wrong is impossible. His sovereignty is His absolute rectitude. His will nothing can constrain or draw aside.

But let me put an illustration of yours in my own way. The sun is revolving to us, but we think of it as at rest, and of man as revolving: Nature is inert, or dead, to us, why should we not think of it as active or living (spiritual), and of man as wanting life? We are not conscious of man's motion, we infer it from his perception: so we are not conscious of man's deadness, we infer it from his perception. And in each case we find a practical benefit in our better knowledge.

W. Thank you.

R. Let me try again. As, in respect to any solid body, that which we can perceive by sight (or see) must differ from the truth of that object by defect; that is, the object must be more than can be perceived by sight—as that which can be 'seen' is only surface or appearance, and to say of *that* 'it is,' would be to affirm the existence of an appearance:—so, in respect to true existence, that which we can perceive by the intellect (or think) must differ by defect from that which IS. That which can be thought is only a phenomenon: and to say of that 'IT IS,' is to affirm the existence of a phenomenon.

Thus, for example, we must think of the world as material, must conceive matter and force; but we should never think of saying it IS so. Matter and force differ from the true BEING of nature by defect.

W. You understand that these are used as illustrations of a proposition independently proved, and not as themselves proofs.

R. I understand. That which is not the absolute existence of nature must differ from that Absolute in not acting. Therefore, if we do not know the Absolute, the phenomenon, which we do know, must be inert. Let me ask you a question here. The phenomenon may be defined as that which is present to consciousness. Now when that which is present to consciousness differs from that which truly exists, of course it will be inert; it will be a phenomenon only, and cannot act. But might there not be a case in which the phenomenon and the true existence should be the same? Might not that which is present to consciousness correspond truly to the Absolute? And then would not the phenomenon also be not-inert, also be active or spiritual?

W. I think with you; in such a case, in which the phenomenal and the absolute should be one, or perception agree with the truth of things, the Being would surely be *consciously* in a spiritual world.

R. Now I see that I have been defining the very thing that you call the true life of Man. If Man's defect or deadness were removed, then that which he is conscious of would not any more be defective or inert. The true existence of nature would be to him as it is; he would perceive things as they are, and the distinction between phenomenal and absolute would be done away.

W. It might be so. This would surely answer well to the idea of being in a spiritual world, that the world should be to us, as it is in truth, spiritual; and this could only be by the taking away from us of the defectiveness which modifies our feeling. That would be to escape from the physical. The thought seems consistent and simple enough; yet I would not affirm it. These simple and natural thoughts are apt to deceive. They must be false when our knowledge is imperfect, and I do not think we yet know all that must determine the answer to your question. Perhaps on the other hand, the phenomenon may always, and necessarily, differ from the Absolute, but when man's defect of being is removed, they may be consciously associated in his feeling; and the physical may present itself to us aright, not as that which IS, but as the mode under which that which IS appears to us. This latter thought more commends itself to the affections. Do you see that so this beautiful phenomenal universe (it may be under ever changing forms) might always remain to us, but not as the reality; being known and felt as it is, in its true dependence on spiritual being, and as perceived by us only by our relation to that being? Do not you see how in that way we might retain our hold on all that we have loved and lived for, not losing it in passing to a different state, but having it glorified and gladdened, and enriched with unspeakable meaning? and not only so, but we may conceive of unnumbered universes, equal to, or surpassing this in glory, though this surpasses infinitely all our thought. Unnumbered universes, perceived, enjoyed, and known as the phenomena of the one true, absolute Universe that IS. Thus we multiply creation, infinitely, to our thought. Nor let it be said: that were to make a mere illusion of all. How should it be so? What value is there in *matter?* Will any one pretend that the dead substratum has any worth, of any sort, real or imaginary, and that the Universe we know would not be more rather than less, if the inference of matter were proved a mistake?

R. We have an idea that God's power is displayed, or proved, by the creation of matter: bringing into existence that which did not exist. So we have a certain momentary reluctance to part with it, until we reflect that we give it up only for a Universe more noble, and in which God's creative glory is therefore more displayed. We must be rather glad than sorry, to understand that this low, evil thing is not that which God has created, but that which man has invented; that it measures man's power of apprehending, not God's power of making. But I do not know that I wholly see your meaning about unnumbered universes, perceived by different classes of Beings, as phenomena of one and the same true Universe.

W. Will you allow me another illustration?

R. You seem never at a loss for those.

W. They crowd upon me. The relation between the phenomenal and the absolute seems to me exactly to resemble that between false appearances and the truth of things. I should say: as the appearance is to the phenomenon, so is the phenomenon to the Absolute. From an appearance, by considering our phenomenal relations, we learn the phenomenon, or that which is true to thought: from the phenomenon, by considering man's absolute relations, we learn the Absolute, or that which is the very truth of being. The one process is ever available to illustrate the other.

Thus, for the point in question. Should we not be sorry if, in learning that the stars constitute a boundless universe instead of a limited sphere as they are to us, we lost our old familiar heavens, and no more saw the accustomed constellations? that is, if the appearance were altered by our true apprehension of the phenomenon. And why, in gaining a true apprehension of the absolute, should the familiar phenomenon be taken from us? Is it not enough for us to *conceive* the stars aright? Do we wish them to look different? So, would it not be enough for us to feel and know the universe as it is; why should the phenomenon be altered? Only let it be *but* phenomenon to our experience. Let our Life be in the Absolute, even as our *thought* is of the genuine stars, and not of globelets rolling round on wheels.

As for the different phenomenal universes perceived by different Beings, this will illustrate what I mean. Is there not a different apparent universe to the dwellers on every separate star or planet, if there be any such? Is there not a different appearance of the heavens to every differently organized eye? And yet but one phenomenon:—the stars that we conceive.

R. I must ask you one thing more on this subject: what happens at the death of the body?

W. I decline the question. I avoid expressing

any opinion on that subject intentionally, in order that my argument may not be embarrassed by any mistake I might fall into.

R. But you must have some opinion.

W. Certain things I think: for instance, that men do not pass into the spiritual world thereby, because they are in it now: that they do not come to the end of a probation for eternity, because I find that idea to be a human doctrine, and as it seems to me a mistaken one: that there is nothing in that change to remove the defect under which men are, and which causes them to feel inertness without them. I see in it nothing to make men good, nothing to make them worse. In fine, I am perfectly content to wait for better ground of judging, which I believe will hereafter be found. Why should I be in haste? Do not I know that man is redeemed, that ALL MEN shall be brought to Christ? I have, however, certain individual impressions, which are of purely private interest, and I gather also, from the New Testament, intimations of an anticipation on the part of Christian men of being present with the Lord on the occurrence of that event.

R. I am prepared now to enter into your meaning. Of course, regarding this physical condition as you do, the dying of the body is a different thing from that which it is on an opposite conception. On your view there is no reason for entertaining a positive opinion on that subject. The change is a phenomenal one. On the view that man *is* a spirit in a body, instead of only seeming to be so, and that the spirit is set free at death and enters into the spiritual world, of course men are bound to give an account of what becomes of it.

W. I cannot cease to be astonished, when I think that the entire religious opinions of so many men are based upon their supposition of what happens at the death of the body. I know how natural that feeling is, but if we ask ourselves whether we really do know, surely we must at once admit that we do not. We have in fact adopted ancient heathen speculations, and grafted them upon the Bible.

R. Do you think then that the Bible is silent on the subject?

W. I do not. I think much may be gathered from its words, and if there were any practical necessity for deciding, we might enter into the discussion. But I do not feel it an urgent question. For all practical ends I know enough: I know the redemption. That I believe, not because I understand how it is to take place, but because it is expressly revealed; and because, reading nature by the light of the Gospel, I see it there also.

R. Then in order to believe that man is to be saved, we need not know what happens at death?

W. Clearly not. And further, I believe that we cannot rightly, or wisely, attempt to unravel that interesting problem, until we can feel thus calmly respecting it. The first condition for any true knowledge on that question seems to me that we should be content to wait for it, and be patient.

R. It is a new idea; yet when we reflect, what is there in bodily dying which should have made us think it so great and decisive a change? Surely we have been carried away by the undue influence of the senses, in thinking of it as we have done.

W. I believe that is the secret of it. To sense, bodily death seems a consummation, an ending, a great and terrible catastrophe. It is no wonder that men should have associated religious ideas with it, as they have done.

R. We have distinguished dying from the rest of our experience, as if it had some special connexion with, and determination of, our spiritual destiny. Can that be a mistake?

W. Think of the New Testament. Are there, in any part of it, any such exhortations to secure salvation while life lasts, as are so abundant in our discourses and books of religion? Is there not an utter dissimilarity of tone between the scriptural writers and those who, in these days, urge men to repent on the score of the nearness of death? I know no greater contrast anywhere.

R. We need not pursue this subject. I perceive that this is but part of the general idea. If we have been mistaken, as you affirm, respecting man's present deadness, it is easy to understand that we have been mistaken respecting the spiritual bearing of bodily death. Whether there be any passages of Scripture which imply the ordinary conception of the results of dying, I will reserve for thought.

W. There is one that will readily occur to you: the parable of Dives and Lazarus.

R. How do you understand that?

W. It is a passage I do not profess perfectly to understand. But I think you will see that it does not decide anything in the present question. The expressions which dwell in our minds do not contain the ideas we associate with them.

R. The rich man is tormented after death.

W. I do not deny that evil-doers have this destiny.

R. But between him and Lazarus there is a great gulf fixed, which neither can pass.

W. There is. Is not this the doctrine of the New Testament throughout, and of the Old also? 'Can the Ethiopian change his skin or the leopard his spots?' Does the bad man ever become good by his own self act, is he not always renewed by God? And do not we ourselves speak in the same way? Is there not an impassable gulf fixed by God between the good and the evil? In short, I think the ideas we connect with this passage are not in the words themselves, but are attached to them by us because they are previously in our minds. The words agree perfectly with the repeated and direct affirmations of the absolute redemption. What can have made us subordinate the explicit statements which assert the redemption of all to phrases which form part of the machinery of a parable? Is this in conformity with our own established principles? Is there not proof here that we have brought ideas of our own to the New Testament?

DIALOGUE III.

R. Let us pass to another subject. In respect to Matter, you set aside any authority of our supposed intuitions and ask: What has made it necessary for men to infer it? And you answer, that it is a defectiveness of their own being which has made them feel as reality that which is but phenomenal. Hence, inasmuch as a phenomenon of course cannot act, they have been compelled to infer an unacting substratum. It is a false inference necessitated by man's own condition, and only to be escaped from through better knowledge: in this respect being like all the other false inferences men have been compelled to make. And this is why the question of matter has been so contended. It has been an inference at once necessary and false. However easy to disprove, still while the necessity of inferring it remains, through overlooking man's defect, it holds its ground. Thus comes the state of things which has been so often noticed, that men continue to believe in matter though they admit the arguments against it. That result is involved in the nature of the case. And the supposition of an authority in our perceptions, to vouch without investigation for the true nature of that which causes them, follows as a natural attempt to bridge over this difficulty, until the solution of it be found in man's defective apprehension. But this supposition is opposed to experience, which shows that our direct perceptions are not authoritative as to the nature of that which causes them in the case of individual phenomena :—opposed to reason, which shows that what we are immediately conscious of must depend in part on what we are, and how we are related to the object which acts on us : — opposed to the emotions, which reject with indignant scorn the idea that Nature can truly be what the matter and motion hypothesis represents it :—opposed, in fine, to every sound method of judging of the relation of causes and effects ; for if we grant nature to be matter and force, how can it possibly make us perceive what we do, or indeed perceive anything at all? That beautiful problem, of the relation between the percipient consciousness and the world, has changed itself, under this method of taking our impressions for granted, into the blackest, dreariest, most impassable of gulfs. We come to a sudden halt. Between matter out there, and my sight of a flower, let no rash mortal presume to indicate the least dream of a rational connexion.

W. Yet I thought matter had been inferred in order to account for our perceptions. Why then does it fail exactly when we come to the whole final cause of its supposition? Why not, in any other case, argue in the same way, and in respect to anything which we naturally suppose, but the existence of which could not account for our impressions, maintain their truth and say: it is a mystery. For example, why not so meet the argument that the stars cannot be little white flames? Why not say: they are so, but we must not ask how we can perceive them so far off? Were not that as reasonable as to say, that Nature is an inert existence, as it is to our impression; and when it is argued that an inert existence could not cause us to perceive, reply: that we must not ask such questions; that is sacred? Plainly we are on the wrong path here. Whatever may be the truth, this idea of a matter-substratum has had its day. No-theory of the world, a candid confession that we cannot account for our experience at all, would at least be better than that. Better positively: it would be truthful, genuine, manful; the attitude which a genuine man naturally takes towards that which he does not understand:—infinitely better negatively, for it would leave the path open for a more hopeful and more humble way of attempting the solution. I see no one purpose that the matter hypothesis answers, but that of puffing us up with a vain conceit of knowing all about the world, and preventing us from investigating. That matter-world has just solidity enough to block up our way, and no more. Suppose we had to eat 'matter' instead of meat, or sit on 'matter' instead of chairs, we should find it unsolid enough then I fear.

R. That were Berkeley's theory realized:—men living in a world that is an idea. The very thought of such an entertainment gives a foretaste of emptiness, and a feeling that recalls Satan's fall through chaos. In such case we should be thankful even for the cloud whose presence there we have so much cause to rue. In truth, there always was something ludicrous to an imagination not duly broken in to reverence, in that *substratum* which was, and was not, what we see and touch, and had to be at once distinguished from, and identified with, all things; in which all that we properly perceive inheres, 'stuck in it,' as Coleridge says, 'like pins in a pincushion.' In fact it is on an emptiness that the notion of matter is based; reminding us of the man who pored in vain over the cane-bottomed chair to think, 'who could have taken all those holes and put the cane around them.' But are we not acting rather like the savage, who begins to kick his idol when he has discovered that it is not a god after all? If matter play little part in nature, it has played a great part in human thought.

W. True. That necessary inference, or belief, of matter brings home to man a proof, from which he cannot escape, of what he is. Be the truth of nature what it may, it has been necessary *to him* to infer an inert, a dead substratum; he has necessarily ascribed to it a being which is itself a denial, an essence which insists on being defined by negatives. To him there is defect in the universe, a void and darkness. He cannot deny that that which he consciously perceives is inert. Matter witnesses against him.

R. So you would meet any one who should deny that what we are conscious of perceiving is inert, and thus avoid the conclusion of man's deadness.

W. I should point him to the matter-hypothesis, and ask: what does it mean that man should have been compelled to infer an universal not-action, and therewith a not-acting substance, in nature?

R. That a man should have been compelled to infer an universal not-light in nature would mean that he was blind.

W. Does it alter the case that all men are included in it?

R. You deny an inert world. Very good. You do not deny that there *is* a world, but that it is inert. Show us, then a reason why we feel it as inert when it is not. It is simple enough.

W. So also is the question whether there is matter: it is simply whether man has been mistaken in thinking the world inert; whether he has erred in interpreting his impressions? We have given up, as not pertaining to that which *exists*, every property that we are conscious of perceiving, except *shape*. Yet what reason is there for retaining that, or what advantage in doing so? Something different from colour, sound, temperature, odour, causes us to perceive all these things; then why not something different from shape cause us to perceive shape? or something different from hardness cause us to perceive hardness? Why should not our thoughts be raised to a consistency among themselves? Why inharmonious laws to regulate our perception? especially when we know so well that shape is not always perceived as it is, but varies continually. If we are caused to perceive a certain shape by that which is different in shape, why may we not be caused to perceive shape altogether by something different? And as for resistance, the feeling of hardness is evidently dependent upon conditions of ourselves.

R. Few things would be hard to us, if our fingers were of sharpened steel, and our muscles iron and steam. And again, it has been well remarked that if our fingers were magnetic, and a central point between them were of like magnetic character so as to repel them, our feeling must be the same as in grasping an elastic body.

W. In short, we ought to have outgrown the assumption that the cause of our impressions is something corresponding to them. It could not be so, and ought not: if it were, how could we obtain a knowledge of ourselves? We deprive our consciousness of more than half its value, by maintaining the existence of that which corresponds to it. We rob it of its power to indicate, by its modifications, our own condition. God gives us in our consciousness an opportunity of learning two things:—that which exists, and our own state: we would indolently assume the former, and ignore the latter. But in repudiating the one, we fail also of the other. For, not caring to inquire what in our own condition *modifies* our consciousness, we merely delude ourselves with the persuasion that we know what *causes* it.

R. But physical things have a certain reality. You must not deny this. You must reconcile your idea with the feelings of mankind, or it is not yet perfect.

W. I agree with you. But I think the reconciliation is already effected. I affirm, in the true sense, the reality of these things that we perceive; for how does it make a thing less real to us, to know that the cause of our perceiving it is something not corresponding to that which we are conscious of? Is fire less *hot*, or are leaves less *green*, because we hold that we are caused to perceive heat and colour by something wholly unlike them? Are heat and light therefore less real? Why then should fire or leaves be less real, if we are caused to perceive them by something different? We are made to perceive physical things by a true actual Existence apart from us, of which they give us sure demonstration. This is what men affirm: there is some real existence acting upon me to make me perceive, and through my perceiving I can be sure it IS.

R. Since that which truly acts upon us, to make us perceive, is by common consent above our comprehension, evidently it follows that what we are conscious of, or do comprehend, is not that very existence itself, but is a phenomenon. Now if we ask respecting a phenomenon whether it exists, in the sense of truly and absolutely existing, evidently it does not: such existence is not an appropriate idea to apply to it. It exists as a phenomenon; it has all the existence which belongs to its nature, but its nature is such that true existence could only be absurdly spoken of in relation to it: that would contradict our own definition. But that which we thus know to be but phenomenal is real to our experience, thus proving defect in man: a defect to recognise which, and trace its effects, is the key to human life. Our error has been of the kind called, by writers on logic, that of *mal-observation*. All the circumstances belonging to the case have not been taken into account; regard not having been paid to man's own condition in interpreting his consciousness. This is your view.

W. And what do you say to it?

R. I accept it. It is but applying to the whole experience of man the same rules that we apply to our own experience as individuals; so giving consistency to our thoughts, and introducing unity into our mental life. The only thing that appears to make it difficult is the novelty. One is apt to imagine, on the first hearing, that the existence of things is called in question; or that you deny the reality of that which makes us perceive. When it becomes clear that our perception, and the certain existence of a reality which is its cause and object, is the basis of the entire argument as it is in the ordinary view, only with a reference to our own condition, which that view ignores or omits; then there is no more difficulty. We ask ourselves as before: Why do we perceive these things? and we answer—not, as before, the action on us of things corresponding to the impressions we are immediately conscious of—but the action of things more excellent than they, which impress us as they do by virtue of man's condition. We do herein only what we have all our lives been learning to do, and are accustomed to do in every single case. We are no more at a loss to think how any particular object which we perceive should not truly exist as such an object, than we are to remember that the bright spot we see as Venus does not exist as such a spot. That which exists is different; but these objects are to man. Their relative existence remains the same: it is not as if we left all the other specks in the heavens as existing, and denied one of them; but we affirm of each one that, in truth, it is part of a whole which is different from that of which we have the impression. When a person looks at the stars at night, he would not deny that there are specks in the heavens. Those specks have the existence which belongs to them. They exist to our sense. But if we ask what is the true existence indicated by them, that is another question: we must take into consideration man's condition, his present mode of being.

W. Exactly so. Thus 'matter' is not the unknown existence of nature, but the known phenomenon. It is that which is to us; having the subjective element in it, as the metaphysicians say; which is, simply enough, defect. Not that we have truly added anything, but that we have not adequately apprehended.

R. That will do. The idea of matter is the attempt to conceive an existence, or substance, corresponding to a defective apprehension of that which exists: necessarily, therefore, a defective substance; necessarily, therefore, a cause of perplexity, to be escaped from only by remembering that our apprehension *is* defective, and that there is not anything corresponding to it, but only something excelling it. For matter, being a substance answering to our inadequate apprehension, necessarily is insufficient, not equal to that which nature is. Thus we have had to suppose also those marvellous powers in ourselves, which convert mere material and mechanical processes into our exquisite sensations. We have had to supplement the palpable insufficiency of the substance we have supposed for nature, by gifts of our own. And the whole is a mystery not to be inquired into. I begin to share your impatience with that plan, not of cutting, but of tying knots. For who is to judge what mysteries are to be allowed, and what are not? What scheme of false assumption might not shelter itself under the same claim?

W. We need never take this position, of predicting the future direction of human thought, if we will allow that our way of regarding its problems may not be perfect. We may safely say of certain questions that they cannot be answered; but we can never say that they may not be found to be mistaken questions, and that the true problems at issue may be solved another way. To take this very case of our sensations as an instance: doubtless, no man will ever be able to say how matter and motion can make us perceive light; but it may be possible for men, hereafter, to say why our perceiving light made us necessarily infer matter and motion; and also to discover what is the true cause of our perceiving light. A defective feeling on man's part already gives a partial answer to the former question; the latter waits. We find an insoluble mystery here, only because we assume as certain that the operation of matter is the cause of our perceiving light. We put an inference, necessary to us in ignorance, as if it were a thing ascertained and known. In a word, we deal by the phenomenal as if it were the absolute.

R. Let me continue. We know as a fact that men sometimes feel that to be which is not. We do so in dreams. Nay, in every case of insufficient knowledge, in which we are deceived by appearances, the same thing may be said. Every one who has an inaccurate impression respecting things around him may be said, in some sense, to feel that to be which is not. Defect on our part has a false feeling for its necessary consequence. Not mere absence of right apprehension, but positive wrong apprehension, is the necessary result of want of a true appreciation of that with which we have to do. Feeling that to be which is not, is a familiar and well known fact in human experience. It is natural to the present state of man; a normal part of his present training. What more reasonable than that it should afford the explanation of this larger experience, which we call the perception of the material world? Man feels that to be which is not: necessarily and rightly feels so, by reason of defect. It is his work to learn and to escape from the error. Therein is a sign and result of his advance. All his work in gaining knowledge is this same process of escaping from error. A familiar fact of our experience gives the key to man's feeling in respect to the physical world: a known and natural circumstance, proved in ordinary life, applied to a larger problem. So weight, the familiar heaviness of bodies on the earth, gives the solution of the gravitation of the spheres. And the known defect of man gives evidence that it must be so. A defective being ought to feel that to be which is not. Else were all the intellectual laws confounded.

Admitting, then, that in his feeling of this physical (phenomenal) world man feels that to be which is not, are not many things made clear? Apparently hopeless mysteries solved at once? And those intricate questions respecting the absolute and the phenomenal, the temporal and eternal, the realities of the spiritual and the shows and forms of earth, stripped of their darkness? The difference is between that which is felt by man to be, and that which IS.

W. Have we not continually to deal with children on this same principle, teaching them to distinguish between the truth of things, and that which they feel to be? My little boy said to me the other day: 'Papa, the lamp jumps when I jump.' To him there was a jumping lamp, as to us there is a material world. Jumping-ness in the lamp was as much a necessary inference to him, as materialness in the world is to us. He had not learnt to understand that the mode in which we are impressed by things depends upon our own condition. Nor have we yet applied this dearly bought knowledge to the aggregate of man's experience. We puzzle ourselves about the *material* world:—how it can be, what it is, how we can perceive it, how we can harmonize it with our belief respecting God and spiritual things, how we can reconcile its existence, and the things that take place in it, with that which is known in the conscience and the heart:—we strive over this problem just as a child might do, who should strive to reconcile a jumping lamp with the little he might know of physics. We should say to him: Oh child, the lamp is not jumping as it is to you; think of yourself. So we should say to man: Oh man, the world is not material, not dead, as it is to you; think of yourself.

R. Applying one law to the individual and to the race; not thinking that there is a sudden break and disharmony between our experience as individuals, and that of man as man. The universality of human experience—that all men perceive in the same way—shows that a common cause is acting upon them, and that they are partakers of a common nature and condition. But what it is that acts on them they must learn by the joint study, at once of what they perceive, and what their own nature and condition are.

W. And think of those anomalies in our philosophy, dreams and illusions of the sense. If man's experience be as you have said, then dreams are natural, and might have been foretold: they are aids to our thought. Consider how utterly they overthrow the argument from our consciousness, or unavoidable conviction, for true existence in the objects of sense. For at any given moment we have this firm conviction of the reality of perceived things, which is supposed to prove them. But let us imagine that the next minute we have the sensation of waking up from sleep, followed by a consciousness as of different things around us. Should we then think of affirming the reality of those things which we before felt ourselves as perceiving? Should we not say at once and unhesitatingly: I had a dream? There is no possible evidence of sense which would not, under these circumstances, be set down to dreaming. But how can that be valid evidence which can thus become invalid? Does not the existence of dreams utterly overthrow the pretence of such authority in our convictions? There are the convictions without the authority; and occurring in such a way that we must own our convictions might at any moment be shown to be unauthoritative by a change in our sensations.

But again: think how beautifully dreams and illusions illustrate the problem of physical perception. What are they but states of feeling in us, which we irresistibly refer to objects which exist only to our feeling, and attribute to causes which have not *being* enough to cause them? We learn afterwards to look for the true cause in other things. Why cannot we interpret man's experience in physical perception, as we have learnt to interpret dreams? Are we any losers, any the less prepared to act wisely and truly, because we understand that the things we are conscious of in dreams do not truly, but only to our feeling, cause our consciousness? This perception of a dead world, of a world that cannot BE, that lands us in utter doubt when we investigate it, what should it mean but that man has consciousness produced by some true causes, but which he feels as if produced by others, and refers to others which have not *being* enough to cause it? The feeling these physical things to be realities is not the true waking life of Man, even as feeling dreams realities is not the individual's true waking life. Man walketh in a vain show. All are included in it; having a consciousness truly produced by one cause, and referring it to others which are unreal. It is a known case, and no hypothesis.

But we may go farther. Only in dreams and

illusions have we any means of getting behind our consciousness, as it were, of testing it, and ascertaining its true nature and relation to the things consciously perceived. And in this case we find that something different from the objects of our conscious perception is the cause of our perceiving them. Should we not think that the law of our perception is given us in this? especially since it is confirmed by an analogous experience so wide, and a necessity in the nature of things so demonstrable. In dreams, or other illusions, our conscious perception of inert things, being transient and limited to the individual, can be analysed, and its nature demonstrated. But that great consciousness which includes all men cannot be so treated; whatever its cause may be, it does not, in our present experience, cease its operation; and it is not limited to one or a few, so that it can be tested by others. Of its nature we must judge in a different way: not by direct experience, but by evidence and proof. Dreams are, in relation to the universal conscious perception, as the perception of motion in particular objects, through our own individual motion, is to the universal perception of motion in the heavens. In respect to the latter, men have no means of experimentally testing their consciousness; it is the same for all and at all times; but the particular and transient motions of individuals give them the means of interpreting it, as due to a state affecting all men. So do our particular and transient perceptions of physical things in dreams, wherein we know that the true cause is different from that which is consciously present to us, give us the means of interpreting the universal perception of physical things, as due to a different cause operative upon all.

Yet farther: in dreams and such illusions, evidence is given that the workings of man's own structure, so to speak, cause him to have conscious feeling of physical things around him. This is a result to which the internal mechanism works, independently of the cause which puts it into operation. So the mad or intoxicated man, by the state of his own body, is consciously surrounded by, and feels every way as reality, an entire world of his own. We, looking on, perceive that his true surroundings are wholly different from those which he feels, and of the non-reality of which no reasoning, no demonstration, could persuade him. How can we be sure Man is not dreaming? not delirious? Does not every language affirm him so? Do not his actions bespeak it? Does he not evidently act by the world not according to its nature, and fail; though to his own feeling he seems all right? Above all does not the divine book tell him in plain words that very thing?

R. Disease, which is in respect to the individual a defect of life, causes him consciously to perceive, and feel as acting upon him, things that do not truly exist. Here is the exact parallel to that which you assert of man. And do we not say, that one great evil of this present state is that we feel these earthly things as too real, that they have an influence over us which they ought not to have, and would not have if we were right. 'If we could but escape from the influence of matter,' we say, 'from our lower, our physical nature.' We want, not to feel these things real.

W. Thus dreams and illusions are turned from casting doubt on our belief, into supports of it; from being exceptions, into examples of our experience. I conceive no man ever felt that the world was less real, because of dreams and their explanation. So can no one feel that there is less reality in the world that truly is, because our feeling respecting the world of phenomena may be likened to a dream on the part of man.

R. There is a point in respect to dreams which you have not noticed. They seem only to repeat that which has been wakingly perceived, being representations in an unreal mode (felt as real), of that which has been before experienced, though in wholly different relations. So may we not believe that man's feeling of reality in that which is unreal, must involve some prior experience of a true reality, correspondent though differing?

W. I do not venture to speculate so far. I feel, in the certainty that Nature is, truly, not inert but spiritual, such a joy that I fear to peril it by uniting with it things which I do not feel also to be susceptible of proof. But if that idea were correct, it would answer well to thoughts respecting man, not only widely spread among the race, but strongly suggested by some narratives of Scripture.

R. This however is the practical point. Here is the phenomenon which we perceive, and feel to be, —this material world. What then are we to think of that which truly exists, and of ourselves? These questions are two halves, mutually dependent: a true knowledge of that which exists, apart from us, must be gained through a recognition of our own state; a knowledge of ourselves must be gained by investigation of that which we perceive. Surely the means of solving the problem which all man's instincts prompt him to attempt:—what is the very truth of things? are thus placed in his hands. And when, on examination, it is found that this which is perceived and felt by man exhibits a character of defectiveness—a defectiveness recognised at once by the intellect in science, and by the heart in religion,—do we need any other evidence that such defectiveness must be the appearance due to man's imperfect appreciation? Or, if other evidence were needed, is it not more than sufficiently supplied by the reasonings which have demonstrated that this defective world, that is felt by us, cannot be that which truly exists; disproving its existence, and mocking us either with an *idea* for a world, or with the offer of believing in a world as existing, which yet we must grant can be disproved, try to maintain it as we will? How can we accept either of these alternatives? Must we not say, is it not sufficiently proved, that the defective world, that is felt by us as existing, is not the true world that IS. Should we not gladly take the defectiveness for our own, that our Universe may be no more defective; and that its existence may be made capable of true demonstration, not resting on the more than doubtful accuracy of our impressions?

W. But what should you say, if you were met by the argument that we cannot know anything about what truly acts on us, but that man can only know phenomena, and must confine his attention to them: that all other attempts must be vain?

R. I should say, that was a vain argument, being an *à priori* one on a subject which must be studied experimentally.

W. But if it were replied, that the attempt has been made, and has failed, and that it is proved that man can only know phenomena?

R. Then I should say: I will try again upon another plan. I will take into account man's own state as influencing his perceptions; I will remember his proved defect. Acting thus, according to reason and experience, perhaps the result may be different.

W. Suppose you were told that it is merely a matter of words to say that the absolute is spiritual; and that, although it is proved not to be inert as the phenomenon is, it is still unknown as much as before.

R. I should say that it is not so: that true action combined with unchangingness, true action and yet necessary, true action yet in absolute fulfilment of law—that this is as much known to be holiness as anything can be known at all. I know that if the necessity in nature only seem to be inertness, then it is truly rightness. But I cannot believe that any one will use this argument against you. You are not opposed to those who have maintained this position respecting the sphere of human knowledge. To me you seem entirely to embrace and affirm their position; and to carry out to the legitimate conclusion their own premisses.

W. If I understand them rightly, I do so. The true positivist doctrine surely is, that man could only know the Absolute by knowing himself, so as to exclude the subjective element from that which he perceives. No one, so far as I am aware, has adduced any argument to show that this cannot be done. I am satisfied that no one can have any unwillingness that it should be done. Towards this result I seek to take one step; as it seems to me the first, and one essential to any farther advance in that direction. It is very simple. One part of the subjective, or human, element in the phenomenon is its defectiveness: the phenomenal differs from the absolute in its inertness. Surely it is a truism.

R. A truism?

W. Yes, a truism. Do you think I thereby diminish its value? I think, on the contrary, that overlooked or forgotten truisms are among the most important of all things. What is geometry but truisms applied? The certainty of knowledge seems to me to rest upon its being capable of analysis into truisms.

R. Let us not diverge from the subject. Show me the application of yours.

W. Thus. The absolute essence of nature, that which IS, is not inert: the phenomenon, or that which is to man's consciousness, is inert. Therefore man introduces inertness into nature.

R. The whole thing is contained in that. If the true being of nature cannot be inert, then the inertness which characterizes it to man's consciousness must be due to his own state; and the foundation of the doctrine of man's deadness seems firmly laid. To deny nature to be inert is to affirm deadness in man. Upon that link the entire chain hangs; whether that which is the true being of nature can be inert, as that which we are conscious of perceiving is.

W. Can it?

R. It seems to me that it cannot be so. If there is any certainty at all, these two propositions surely possess it. First, that the true existence which acts on us does act: and, secondly, that if we do not know this true existence, then that which we do know cannot act, because it is not that which IS.

W If, then, we know the absolute in nature to be not inert, is not that some knowledge respecting it? Is it not a beginning of knowledge? And if, in attempting to learn the absolute from the phenomenal, we have to remember a deadness in man, may we not begin the investigation with some guidance, with some hope?

R. I do not think any consistent positivist would oppose you there. He would feel that you do not deny his position, but take it as your starting point. Their doctrine is: Let us attend to these phenomenal things till we have better means of knowing. Not affirming, or denying, of the future; but waiting humbly and hopefully for whatever that future may bring.

W. In that I would be one of them. And also, in treating all things with regard to the progress of humanity. The true charm of that system is in its subordinating all private interests to the welfare of man, and treating all material things as of no value in and for themselves, calling them phenomenal. The world has vast obligations to Auguste Comte, not the least of which is his scientific demonstration of the authority of the moral over the merely intellectual. And has not that man a title to our love, who does not shrink from ridicule in advocating, according to his best thought, the claims of the affections?

R. It is evident that to regard physical things as only phenomena, and treat the intellect as dealing only with the relative, enthrones the moral at once in indisputable supremacy. That doctrine, in whatever form propounded, must be fundamentally ethical. If it be not so, it is nothing—a science merely of dreams. Of necessity, it concentrates its regard upon the spiritual (for positivism does not disown the word), and upon the spiritual in its bearing on man.

W. Has it not struck you that my practical conclusion is identical, essentially, with that of the positivist? Regard in all things man's raising up:—from his dead state, I say (feeling the words of the New Testament to be the exact words I want)—from his imperfection, says the positivist. I do not shrink from owning the oneness, I rejoice in it.

R. You should do so.

W. And I have larger grounds for being glad. Is there any class of thinking men whom I oppose? Do I not truly agree with all?

R. That has been held impossible. Yet ways of doing such impossibilities may be found.

W. To believe the felt inertness to be due to man is one. That shall set you free from all necessity of contradiction. Once deny that the Universe is truly dead, as man feels it, and you need deny no more. The 'everlasting yes' has made its home upon your lips. For what is there that man affirms that you do not also affirm? to what conviction of his heart, or intellect, or conscience, to what aspiration of his nature, need you turn any longer a deaf ear? Is not the whole course of human thought just that which, on such premisses, it should have been? Are not all its averments embraced alike with love; none rejected, none scorned, but each, with tenderness and reverence, receiving its due weight and honour? Does not the mental life of man point to this issue, claim this victory? Do not extremes meet here? the most rigid logic and the largest love, the steadiest grasp upon experience and the highest raptures of enthusiasm? Is it not the making one of doubt and faith? the giving up of all that is not proved, and the holding fast of all that is to be desired?

R. Do you mean to say it unites opposite doctrines, Positivism for example, and spiritual religion?

W. It unites them, if to unite is to show that they must be one. The 'unknown absolute' of the positivist is the 'spiritual' of the theologian. There is no contradiction between them. The assertion on the one side, and the admission on the other, that that which is consciously perceived (the physical) is phenomenon only, breaks down the wall of partition. These are two halves which wait a predestined union. If this world be but phenomenon, of what is it the phenomenon, but of that spiritual world which our consciences attest? If there be a spiritual world present to us, operative upon us, what does it make us consciously perceive? What is the phenomenon of it, but this physical world, which is known to be the phenomenon of something different from itself? Who should forbid the consummation upon earth of a bridal so long prepared in heaven? For what is needed but the opening, on each side, of a narrowed heart and niggard hand? Accept the result of examination of the world, which says this known physical is not a true existence but an appearance of some other; accept the conscience, which affirms another existence different from the physical; recognise the known defect of man, which must cause the phenomenon of which he is conscious to differ by defect from that which is: and what remains to do?

R. Here, you would say, is a known phenomenon (the physical) inertly necessary, which lacks an absolute; and here, a known absolute (the spiritual) actively necessary, or holy, which lacks a phenomenon. To recognise man's deadness, which makes the active inert to him, brings them into one.

W. And the theologian, however real he may assert this physical to be, holds that it has only an inferior mode of existence to that of the other world which he affirms. This inferior mode of existence the positivist defines, showing it to be phenomenal; that is, an apparent existence only, due to the action on us of a true existence which is different. And both agree that what the common instinct of mankind affirms is this inferior existence of the physical, and no more.

R. It is clear that positivism necessarily asserts the existence of another world than this with which our senses are conversant. That lies at the basis of its entire doctrine. If it were not so, whatever the things we know might be they could not be 'phenomena.'

W. The strife is at an end. The opposites refuse any more to be opposed; they have gravitated into one. The strong attraction of mutual helpfulness and need has overborne prejudice and caprice. Each in the other recognises its second self, the twain have become one; the unknown want interpreting itself in its fulfilment. So youth and maiden dwell in solitude, or meet with jealous pride, until each finds that their want is the other. For does not positivism lay avowed constraint on certain tendencies in man, and bid him hold his longings under rigid check? Does not piety acknowledge not less restraint; confess too often an opposedness in earthly things, distraction rather than help from daily life, a necessity to check the current of spontaneous thought, and leave some inquiries unpursued? Each gives the other liberty. The positivist receives a known for his unknown absolute, on which his pent-up thought may expatiate in freedom: the man of piety casts off the weight of a world of realities opposed to godliness. It has become the very sphere of godliness itself, the nourisher and upholder of his piety, the present proof and evidence of things not seen. To know the physical to be the phenomenon of the spiritual makes Christianity and science one.

R. That were a union man should celebrate with songs, and earth array herself in festal robes to greet. But if the view that deadness is in man and not in nature thus unites extremes, which had no thought save of inextinguishable war, does all that is between find equal reconciliation?

W. Think of the secularist who affirms that this is our only world; that it is our real, and sole concern, and that its laws alone are to be studied. He means, of course, the reality and very fact of it; wherein we do affirm the same as he. The very fact of this world is the real and only fact with which we have concern. Its laws are the only laws for us to know. And the dreaming mystic who opposes him, and says that all this is transient and unreal, and turns aside, neglectful of his daily duties, to a world unseen; do we not embrace him too? And, by affirming more fully his own thought, win him back to an observant thoughtfulness towards things around him, and a practical activity?

Or again, are not the idealist and the asserters of common sense made one? With the former, we admit that sensible things, being phenomena, can exist only in a mind: with the latter, we affirm an absolute existence (not in a mind), as being that which is the true cause and object of perception. We do but carry out and complete that position, excluding from that of which the existence is affirmed the qualities which are made to appear in it by the laws of man's perception. As they exclude colour, for example, so do we exclude inertness: as they say, that which exists is not itself coloured, but has the power of causing man to perceive a coloured object, so do we say, that which exists is not itself inert, but has the power of causing man to perceive an inert object: we recognise that another perceived quality does not belong to that which exists. We are wholly of the school of common sense; yet give our hand to its ancient foe, and claim him as an ally; for when we add inertness to the list of subjective qualities, nature expands and rises so, that we need his aid to grasp the overwhelming truth. Those also we agree with, who affirm that the reason we perceive material things in time and space is, that they are in time and space. That is true. The phenomena are in time and space, and must be so perceived, if perceived as they are. For time and space may well be called *conditions of the phenomenal*: they belong to it, characterize it, are inseparable from it. And he to whom phenomena are realities must be in time and space. He only can be not so, to whom phenomena are but phenomena, and not realities. So we do not oppose those who have affirmed that space and time belong not to that which truly IS, but are conditions only of that which is to us. And with those also we agree, who maintain that our faculties must be trustworthy, and that man 'is not a phantom in a world of phantoms.' He is not. He is a defective being, to whose apprehension and feeling things are not as they truly are; a being wanting in life, amid a living world. And his faculties are exactly what they should be; they teach him this very thing in the best and rightest way. And finally, we must not forget how many men will tell us that we might have spared our proofs that the world is truly spiritual: how many hold that to be self-evident, and treat with scorn the notion of a dead substratum. To these also we tender our allegiance, and confess them right; yet plead with them for recognition of their brother's rightness too, in affirming that a deadness is perceived by man, and that a dead substance must have been inferred.

R. You would make me believe, in spite of myself, that I agree with every one, and that men have not really gone wrong in thinking so diversely as they have. One unrectified mistake has necessitated all; and all contribute their part to the solution. But do you leave no scope at all for the warlike instincts of our nature, nothing for us verily to oppose, and to put down? It were ill done; if you made men to be at peace among themselves, they would fall with unanimous assault on you: the instinct were too strong.

W. There is no fear: scope shall not fail for impulse to strife; nor absence of an enemy baulk the expectant arm. Our zeal may burn against the self in us. Against that foe we may wage a warfare wherein victory will be true victory; a fight that is verily for LIFE.

R. That is true, if self be our defect; for then to have self cast out from us is truly to have life bestowed. But it must be done for us. How can the self cast out the self? How can Satan cast out Satan? Our willingness and our exertion, these must be the grace of God within us. We labour, yet not we. But speaking of our self, it interests me to bring this view of it into connexion with Professor Ferrier's admirable book: '*On Being and Knowing,'* in which he argues so powerfully that in all our perception the self enters as an element, and that, apart from the self, the objects we are conscious of cannot exist, being indeed constituted of the union of self and object. Take the self as defect, a negative element, and this view is the same as yours: a negative element introduced into nature by man's state of being. Thus too the phenomenon is reality to the self.

W. I feel it a great confirmation. If this self be man's defect of being, and his self-consciousness be consciousness of defect, how wide a harmony is in the present, how bright a prospect in the future. All this life is to cast out self from man; to raise him from a consciousness of death to a true living consciousness. To this end all things work, the evil and the good. Thus working, all things are to God's glory; the glory of His giving life to man.

R. To accept the self of which we are conscious as the opposite to being, would make clear many problems. How simple then, that this state is one of deadness; how simple that sin should follow: that God is leading the world aright through this dark labyrinth. How simple that we should be deceived by a false feeling, and find as it were opposing lives within: a self-life, which is truly death, and a true human life, the gift of God. How simple that our best plans and purposes, based on the self, should fail; that all contrivances, which treat this self as if it were man's being, should come to nought: that reason should seem to be baffled here, and man's best judgment leave but a darker mystery, as his best efforts have so often left but deeper evil. To know this self truly, as not humanity but as defect of it, would turn all this to light: thus the Christian code of absolute self-abnegation is the true reason; thus utter loss is perfect gain; thus death is Life indeed.

W. We have taken a self view of the world and of all things, and it will not do. From it comes darkness in the thought, sin in the life. But so it should be. How could the universe be truly good and clear to MAN, if it were not dark and evil to his present self? The mystery and the misery to man, as he is now, are pledge and proof of the goodness that he shall know to be, and ever to have been, when there is no more death in him: of the joy that he shall feel when Life is perfect in him, and in love alone is joy. If we could but open our eyes to see all as it truly is, not as it is to self, heaven were already before us. For is it not a privilege of heaven to be so utterly given up to God, so filled with the feeling of His absolute goodness, and His absolute control, that the gladness of the joy that is in Him rises ever to the fulness of our power to rejoice: leaves us no time nor heart to think of self, nor care for it, because our soul is wholly filled with Him?

DIALOGUE IV.

R. TO recognise self as defect seems to unite departments of thought which have, hitherto, been treated as distinct: those, namely, which relate to 'being' and to character. The question of 'existence' ceases to be a mere speculation; it becomes spiritual, and links itself with our deepest feelings. To BE is to cast out self. We seem to catch a glimpse of a deeper reason and necessity in the moral elements of our nature. Virtue must be in self-control; holiness in self-sacrifice. The moral and spiritual laws are primary: Man's true being and his sacrifice of self are one. Life can only be affirmed where there is holiness. Self-will must be called death: it *is* so.

W. It is most true. Being and holiness are inseparable, for being is spiritual. That abstract conception, which passes in our intellect for 'being,' is not the true being. That is a notion merely, and has only arisen through our feeling the phenomenal as real. There is no such inert existence: it can only seem. BEING is a word of infinite meaning, which refuses to be thought. It carries holiness with it. We are conscious of being evil because we are conscious of death. Perhaps a good thought, for us, is that being is the opposite of the self of which we are conscious, from which the evil comes. So perhaps we may deem of it somewhat more worthily: so think more rightly of God, never dissociating our thought of his 'existence' from moral elements. But to see truly what it is to

BE, I know of but one object on which to fix my eye: on Him I look, who reveals to us God.

R. In Him was Life.

W. And the Life is the Light of men: the light wherein we see what BEING is. It should not seem a strange thing that when God reveals himself to us it is as a sacrifice. If we had had life within us, we should have known that it could be only so: that there is no other way in which Life can be shown to such a self as ours; even as light can be revealed in darkness only as its destruction. To self, BEING is sacrifice.

R. Because 'tis Love. God is not a substance with 'powers' inhering in it, such as all the things we conceive must be. Surely that is spiritual which is Love, in the sense in which God is Love: unthinkable.

W. Say rather, that IS. For that alone is active. Action and love are one: how can Action be except in giving; in the outflowing of the life and power within? To us, who have this consciousness of self, Action must be the giving up of self.

R. So our only true action is in self-sacrifice, in holiness. That is to be truly man. This might be your answer, if it should be urged that man is not *only* defective, but that there is in him positive wickedness.

W. I affirm that also. The two affirmations are truly one. We need only to understand that being is not to be thought; that being is holy; then is defect, or deadness, also unholiness, the opposite to love. When we are speaking of that which truly is, and not merely of that which is felt by us to be but is not, being is a word of spiritual meaning. And defect is self: it brings the sin which is in self-assertion.

R. Sin is involved in the affirmation of deadness, because Life is not a mere passive state, but is holiness?

W. Even so: only that which is but phenomenal is passive. The reality of life is spiritual, as is all true reality.

R. The essential point appears to be the admitted doctrine, that that which IS—or BEING—does not correspond to that which we conceive: or, in other words, that being cannot be thought. We must, therefore, admit it above thought, and recognise in our necessary modes of thinking indications of our own state, not rules by which we may judge of that which IS.

W. The rule I would suggest is very simple. That which exists causes man to be intellectually impressed in certain ways, which depend partly on *its* nature, partly on *his* condition. There is, therefore, no authority in man's intellectual impressions (his necessary, or logical, thoughts) in respect to that which exists; but the problem we have to solve is, in every case, to show how that which exists should cause such thought in man, his own condition being ascertained and allowed for. One application of this principle I have argued for, in respect to our necessary conception of nature as inert; but doubtless it has many others. The intellect is subordinate to the conscience: that which the moral sense affirms, tested and corrected by the intellect as the intellect is by sense, is really true for man; is that which we may rely upon, which will not betray and deceive. And I venture to feel sure, that that which is really true for Man is also really, and absolutely, true. If man have his true Being then he is a standard of Being. He is not so now through his defect, because of the nature of the self that is in him.

R. Man's instinct that he can judge of Being, can measure existence by his own standard, is not in itself a mistaken one; it is vitiated only by his unrecognised defect?

W. That is what I mean. His self—a negation—is mixed with everything he thinks; and thereby his thoughts are falsified. His feeling, that things are and must be as they are to him, is a feeling which rightly pertains to him as man, and will be justified when his deadness is done away.

R. Then things will be to him as they are; and the assertion, 'this must be as it is to us,' or, 'this that is to my consciousness must BE;' will not any more be false. That it is false now proves this not the true human life—not the LIFE of man.

W. Thus it is that the assertion of man's immediate knowledge of the spiritual, of his direct intuition of divine things, fails to gain the assent of men. It is true of man only in his truly living state. It must have been asserted; but ignorance of man's deadness vitiates the statement. Like the assertion of man's freedom, it applies to a manhood which is not ours, to a life which is yet to be bestowed.

R. Can this be the heresy of those who said 'that the Resurrection is already past'? affirming that man has already his true life, is already raised up from his dead state, while this self is yet in him?

W. Does it not seem natural? What other 'Resurrection' could be supposed already past?

R. Can it be that the writers of the New Testament sometimes speak of the Resurrection of the dead in this sense?

W. I wish you would examine. In the meantime, observe how a defect in man explains the apparent inconsistency of telling men not to regard that as real, which yet is real to them: it shows, not only that we may, but that we must, rise above that which has been called the human point of view: that is the self point of view, and it is emphatically not the human. That which is to the self, to the Man is not; and for manhood must be treated so. To say: 'This is to me' does not settle the question; we must analyze this 'me.'

R. Thus when the things which are real to the self, wealth, comfort, honour, all material and intellectual things, are set aside and disregarded, treated as nought, for right or love, then we behold manhood. Then we see men treat things as they are. Those are the heroes, to whom all eyes turn with love, at whose name every heart beats higher. They show to man his manhood, and he recognises it with reverence and joy.

W. They bear witness of the life; but the Power that can make us live must come from a higher than they. He in whom The Son of Man was revealed, who shows us God in very deed, and makes clear this tangled web of earthly life; He who has borne the world's sins, and taken away its guilt; He only, and the Spirit that proceeds from Him and from His Father, is the Lord and Giver of Life.

R. It is no wonder that the world has clung so to His memory; that no abuses practised in His name, nor subtleness of reasoning, could loose the impassioned grasp with which a despairing world holds His Divinity.

W. Nor that bad arguments have been used to justify a faith that could not be renounced.

R. 'Tis not a question of argument or proof. Christ shows us God. The sight eclipses all glories else, and constrains our dazzled eyes. Many things we know and love, but when we think of God we think of Jesus.

W. Of Jesus the Redeemer, who makes us know that we are dead, and gives us life. And so explains all things to us, and reveals creation's secret.

R. We find it difficult to admit a state of consciousness to be one of death. Yet is it not quite natural? What death is surely depends on what life is. If life is conscious life, ought not death to be conscious death? If life involves holiness, ought not death to involve sinfulness? Why should not the only possible opposite to life involve such consciousness of self as ours? Whether things are, or are not, as man feels them, depends on whether there is something wrong with respect to him.

W. Each of us may answer that question for himself.

R. In reference to the ideas now entertained of nature, are we not in the position of maintaining the existence of that which is to thought, while yet we assert that that which exists cannot be thought; asserting, as it were, an inconceivable in that which we conceive?

W. I think this is exactly our embarrassment. We have learnt that the true existence of nature cannot be conceived, yet we cannot give up the existence of that which we conceive. We are thus involved in an obvious contradiction, unless we are content to say that we have not any true knowledge at all, and so give up religion. But is not the reason evident? We ought to come to this difficulty; it is the very thing that compels us to recognise man's deadness:—his realities are not real. For 'tis certain that the things we think are real to us.

R. Our life is a life in phenomena: admit it not man's true life—our self as not man's being—and all is clear.

W. Otherwise only impenetrable confusion, which all attempts to clear up make only the more manifest. But how simple the solution is: it is only to remember that there is defect in man.

R. I have observed the extreme obviousness and simplicity of that which you lay as the foundation. It seems strange that so much should come from merely taking into account, in our thoughts, a fact which we never think of denying.

W. But it is perfectly natural that it should be so. What we do, thereby, is to take another point of view, and all things of course look different.

R. You do but say, also, that there is only one UNIVERSE, in which you cannot be held to affirm a novelty. The word proves that some men have been of your opinion.

W. It is man's native opinion; nor do I know a stronger presumption in favour of any view, than that it shows the hidden truthfulness of words, bringing more clearly to men's thought what they have been unconsciously affirming. Men are only compelled to say there is more than one universe, because they think this universe truly is inert, dead, bad, as they feel it.

R. You come to the old point again: consider yourself.

W. How can I leave it, when it gives me such light and joy, such deliverance from bondage to my own impressions? But I was about to observe, that it seems to be in that which we think, or conceive, that man's defect is especially evident: there his self is brought into clear relief. So that our plan, of trying to hold that as existing which answers to our conceptions, is in some respects the worst of all. It would be, in some sense, better to affirm the existence of that which is to our sensation; of the light and sound, *e.g.*, which our senses feel, rather than of the motion which our thoughts conceive. And we should surely have as good ground for the one, as for the other; wholly subjective as both are.

R. Motion itself, of course, is merely a matter of sensuous perception, as much as light or warmth. We do but substitute an idea derived from one mode of sensuous impression for other similar impressions. We have not anything less subjective in motion than in music.

W. I think not. I did not mean to pay you so ill a compliment as to suppose you unaware of this. But has it never struck you, how the life and being of things are turned out of them by our attempts to grasp them in thought? Take the case of nature, for example: this glorious world, which fills us with such emotions, is such a wonderful existence. We try to think it, to bring it before us in orderly conceptions, to present it clearly to our minds:—What has happened? what mystery, what inverted miracle, what miserable abortion is here? Where has NATURE disappeared, leaving nothing in our thoughts but that ridiculous caput mortuum of matter and motion? But this result is what should be: we were trying to think BEING. Again: take holiness or virtue. Think it: and what is left? a scheme for the greatest good of the greatest number, perhaps. Or yet again:

what word fills us with such a feeling of awe, with such a consciousness of the presence of unspeakable EXISTENCE, as the word Eternal? But when we think it, or try to put it into conceptions, what have we? Nothing but duration without limits, time that does not end; no Existence at all, but a mode only, and one which must be defined by negatives.—Why should we perplex ourselves so vainly? Do we not know that we cannot think that which IS?

R. We discover that we cannot, by feeling that it escapes us when we try.

W. Then by what we are obliged to think we know ourselves. Our being such as we are makes the eternal, to our thought, nothing but an everlasting emptiness. But in truth we know the eternal better, even as we know nature better, and holiness better. When love is perfect, and drives away every thought but of the object loved, all reference to reward, all regard to anything to be obtained; when the soul is wholly satisfied in self-surrender, this *now* of utter and final loss being enough of time; when in love the whole life is gathered up into the moment, and it is in this present giving, though it be the giving up of all which leaves no future, that the life is found; that reveals to us Eternity. That makes us know what the Eternal Life may be, when time is not.

R. But we must think of the future, and ask what shall we have? It is our nature.

W. It is. Have we not been told that men are dead? It belongs to this self that is in us to do so.

R. Then we must also ask that question in religion. We must look forward to a future, and consider what *our* lot shall be.

W. True. And the question is answered for us.

We shall be made alive: freed from the necessity of asking that question any more.

R. But Christ had respect to the reward set before Him.

W. And what is it, but that of saving man?

R. That also must be ours. We are to enter into the joy of our Lord: that which is Christ's joy also to be ours.

W. Awe and gladness struggle on our tongues. How should we speak, yet how be silent? Heaven grows so beautiful, and so near. It is no more afar off, but now. Now we have Christ's joy, the joy of man's redemption, and our part in it.

R. I see heaven; and at the same time see why earth must be what it is. For if the joy of heaven is in Love, in giving, in the utter sacrifice of all that is to self, then it is also now. Then are we in heaven and know it not, then are we in heaven and turn it into hell.

W. Because we love the self, which God hates.

R. Because we have not life, and do not know. Man would not be so foolish if he knew. If he knew, in very deed, what God is, and what man, and what it is to BE.

W. Therefore does God reveal Himself.

R. He reveals Himself as the bearer of the sins of the world, as the sacrifice for transgression, as the utter giver-up, to be one with whom is wholly to deny the self; to show us, so, the fact of Being. That is what He IS.

W. It is what He IS. All life is revealed to us in that revelation to us of God. Heaven's light has broken through earth's darkness. To live is to bear, to give, to be a willing sacrifice: to be dead—let us not speak of it, it does not need to be described.

R. Love shall never cease. Only our wrong feeling makes this difficult to us. If we could think of ourselves so changed, that what is sacrifice to us should be perfect joy, that loss should be no loss because there would be no wish to gain, that giving up and sacrificing should be perfect happiness, and nothing else even seem like happiness to us, then we might understand it. Is it that you mean by the self being destroyed?

W. That is it. But remember, we cannot conceive it. Do not try to think it. That makes the eternal an everlasting time. These are things that man conceives not, but God reveals by His Spirit.

R. So these physical things, which are real to us now, and necessitate our self-regard, the conditions that make it impossible for us to conceive that state of absolute self-abnegation, shall be no more real to us. Then man shall not be any more in a physical world, but in the spiritual?

W. I look upon it so. I regard spiritual, and not merely physical, changes as determining the state of man.

R. I am glad our conversation has taken this course that because it has removed, without my expressly stating them, two feelings which were, perhaps, the chief obstacles to my acceptance of your views. The first of these was, that your representation of the work of Christ seemed to exclude the idea of expiation. You dwelt so much upon Christ's revealing God, that it seemed as if you overlooked his sacrificial work.

W. You see that I do not. It is as a sacrifice for us that Christ reveals God.

R. Do you leave the views commonly called evangelical quite untouched?

W. I deny no part of them; but I feel their power and extent increased. And this in two respects: for while I assert the absolute justice of God, and His punishment of sin by the infliction of suffering, I think more of Christ's saving us from damnation, or from being wicked. And surely this is to make stronger the ground of our love to Him. For is not being wicked a worse thing than suffering? and do we not love a Saviour in proportion to the evil of that from which he sacrifices Himself to save us? We must love Christ more for giving Himself to save us from sin, than from suffering, because it is a worse evil. To put most prominently forward Christ's saving us from suffering, as the ground of love to Him, is to invert the laws of human nature. It is never done in the New Testament.

R. That is true. I have often noticed that Christ is always said to redeem us from sin, from iniquity, from vain conversation, from this present evil world.

W. And again: if Christ by His sacrifice saves the whole world, must we not love Him more for that than for saving only a part? It is impossible that our love should not be greater to Christ for saving all men, than for saving ourselves alone, or ourselves and some others only. However much we may love Him as our own Saviour, how can we help loving Him more as the Saviour of the whole world?

R. It increases the power of the evangelical motives to believe the absolute salvation?

W. Infinitely. It renders them unutterably vast; leaving nothing more to be desired, or conceived, of good.

R. But that is only if we see that men are now dead, are now in hell, and damned.

W. True. It demands that we should not banish the eternal into the future, and should admit that to like sin is worse than to suffer. Surely it is not a large demand.

R. It is not a large demand if we can give up our natural impressions, and admit that they may not be true. But the other feeling, to which I referred as being removed without my stating it, was this: that you had constructed a system of philosophy, and adapted the words of Scripture to it.

W. A jealousy in that respect has ample justification. Try me by the words of Scripture, and see if it be so.

R. I have said that the impression is already removed. Yet it seems strange that if this be true it should be so new.

W. It is not strange.

R. Not strange, I grant, if we look at the history of the world, and the progress of human thought from ignorance to knowledge; but it is strange that the very Gospel should come to us as something new.

W. It does not come as something new.

R. Do not be captious on my words. These thoughts are to us both new and strange.

W. I cannot permit you to say so. Look into your own heart, and tell me again if they be so.

R. If you mean that men have always felt in their hearts that to be sinful is the worst of all things; that the redemption they most need is a redemption from the evil of their own nature; that the eternal things are the realities with which they have now the true concern; this of course is not new.

W. What else should I mean? what else have I been saying? Do not you see that I have but translated the everlasting language of the heart into that of the intellect; that I have only laid aside conceptions of the thought which crushed and overbore the convictions of the man?

R. That certain intellectual notions, unavoidable in our imperfect knowledge, have been fighting against that which we feel to be true, and have made a struggle in our life which may be put to rest by taking a truer intellectual point of view?

W. You have said all. I have nothing to add, except that these intellectual notions have done as much violence to the words of the New Testament as to the nature of man.

R. Can that be the reason there has been so much difficulty in understanding it, and reconciling its various passages? that it has been so hard to gather a uniform and consistent meaning from the whole?

W. Do not let this question cease to be asked till you have fairly answered it. Meanwhile let me suggest my answer: we have had a philosophy which has prevented us from believing the Gospel.

R. What do you mean by the Gospel?

W. That Christ redeems the world.

R. But men become worse by sinful deeds, become more and more dead to good, and are not saved.

W. I believe the Gospel none the less.

R. But ought we not to see things otherwise, if it were true that all men shall be drawn to Christ, subdued by Him, in that sense?

W. Ought we to walk by sight or by faith?

R. Do you mean to say that we must lay aside all these natural feelings, and rest simply upon those declarations?

W. If you claim to believe in Christ, surely you

should not find that hard. But if it be, see what helps are given to a feeble faith. Only understand that to redeem man is to alter his nature and Being, to make him new, and then there is no more any difficulty. Sin exists to this end. Because man *is* wrong, and has to be made right, therefore he sins and goes from worse to worse. How should he be brought to self sacrifice in any other way than by learning how evil his self is?

R. But do all the passages in the New Testament agree with this view?

W. Do they not? Does not the whole book become plain and clear? Does not new light break in upon every part?

R. These are questions not to be answered now. If we could see that the saving of the whole world from a state of death was truly the doctrine of the New Testament, how gladly we should believe it.

W. It must be believed as soon as the opinions which make it seem opposed to the conscience are removed: as soon as we can see that to like sin is infinitely bad, and that the bliss of heaven is in sacrifice of self. The heart of man cries out for it, and refuses to be comforted. This is the reason I have sought to show that science proves man's deadness. It is the false opinion of his life, making us think this state his probation, which alone binds our hands from grasping the gift of God, our ears from crediting the glad news of salvation.

R. But if the absolute redemption of man be the truth, why has it been so long unknown, so long rejected?

W. Even that question may be answered. It could not have been otherwise. Had not man to learn nature? had he not to discover the deadness in himself which makes the Universe dead to him?

What other course could he have gone through than this that has been? Does not St. Paul affirm the necessity of a falling away?

R. I am tempted to say, that can apply only to times that are past, but I am silenced by the thought that I should only assert that *I* could not be deceived.

W. Is the world now so good, religion so triumphant, piety so pure and living, that we, of all ages, alone may say, we cannot be mistaken?

R. If it be true that the world is redeemed, there is no more such mystery in God permitting error; even error that might embrace ourselves. It no more involves consequences which we cannot face.

W. True. To believe the absolute redemption gives us patience, not only for ourselves but for the world. Patience is one of the fruits of faith.

R. God does suffer error, inconceivable, unutterable, long enduring. We cannot shut our eyes to facts. It is part of God's discipline of the world that men should err: should feel sure they have His truth when they are upholding their own thoughts.

W. Is not this, in brief, the history of Christianity: Christ and the Apostles proclaimed the absolute redemption of the world, conjoining it with that firm assertion of death, and condemnation, and judgment, from which it is inseparable. But when the first Christians, who had received this belief mixed with much ignorance, came in contact with the world, with philosophy, they lost it. They found men busy with the question of good and evil; and the Church fell as Man had done. They gave up the Gospel, the death of all and the absolute redemption of all through believing in

Christ, and took in its stead the natural opinion, the philosophical doctrine, of the future well-being of a part and the ruin of the rest; accommodating to it the Christian doctrine of faith, in the best way they could. That is the death of Christianity. The conscience was the chief agent in it. The Christians re-adopted the heathen view, of man's life and probation, instead of the Christian one of his death and salvation. The difference between these lies fundamentally in the conception of humanity. The christian Manhood is different from the heathen: the one accepts this; the other asserts a higher.

R. Then you do not reject the everlasting punishment of men on the ground that it is opposed to reason?

W. No. I could not. I hold that doctrine to rest wholly on human opinion. It is a result which flows from our natural impressions, and has been imposed on the New Testament thereby: we see it there because it is in our thoughts. How could I deny it to be a natural opinion of mankind when it is stated most explicitly in heathen writings anterior to the New Testament?

R. It is Plato's doctrine. The everlasting misery of the worst part of mankind is clearly set forth in the *Phædo*.

W. Nor do I see any escape from that opinion, except by accepting the scriptural representation of man. My hope for the world is in that which is written, of 'God our Saviour, who will have all men to be saved, and come to a knowledge of the truth.' Does God will that which shall not be? And besides, I see in Christ a revelation which makes me know that this is man's death, a state from which he is to be raised, not an experiment.

R. But why do you lay such stress upon the saving of all men? You speak as if, supposing that were not, there were nothing.

W. I feel so. It is that which saves me. The knowledge of that is the turning point of life and death, of the possibility of absolute self-abnegation.

R. Is it so great a difference?

W. So great? Does not your heart bound with exultation at the thought that you might, doing no violence to any principle you have learnt to revere, giving up nothing of reverence for justice, of hatred to sin, of value for the Saviour's sacrifice, believe the salvation of the world? In knowing that, a burden as great as that of our own sins is taken from our hearts, a joy-giving and sanctifying power, not less than that of our own forgiveness, diffuses itself through the soul, and makes itself felt in the life.

R. Let me put you on your guard against a misapprehension to which you might be exposed. Do you mean to imply that the absolute redemption of men can be believed only on condition of accepting your opinions respecting the physical world?

W. Emphatically I do not. If I might be so misunderstood I thank you for remarking it. The absolute redemption is to be believed wholly on the ground that the New Testament affirms it, and may equally be believed by all. But I think that the view I have suggested of nature is essentially a christian one. If science teaches us man's deadness, how can we but see in Christ a Redeemer from death? I think that false opinions respecting the world have prevented us from accepting the plain statements of the New Testament, therefore I have brought my attempt to show those opinions to be false into connexion with those statements.

I seek only to rectify a connexion which already exists, not to establish a new one. What I affirm is that men do read the New Testament according to their opinions respecting the world, and that they misread it because those opinions are not true. With better theoretical opinions they would cease to coerce its words.

R. There are innumerable other questions I might ask you:—What is the exact meaning when you speak of deadness in man? how, then, are we to consider man as existing? what relations does he bear to the other existence in the universe? why should there be death in him, and how can it be without infringing on the perfect goodness of the world?

W. These questions are most legitimate; but I cannot enter on them. They belong to another sphere of discussion. Even if I granted (which I do not) that no answer could as yet be given to them, I could not admit that the argument I have carried on would suffer in the least. I have undertaken to give evidence of a fact—deadness in man—the explanation, or reason of it is wholly another question. I cannot consent to mix up the two; nor to make any thoughts I may entertain respecting the latter a mere pendant to the discussion of the former. First let us determine whether it is true that man's deadness (or defect) is the cause of his feeling the universe to be dead (or defective). If we answer that question in the affirmative, a new inquiry lies before us:—What is the meaning, what is the cause, what are the relations, of this fact? I should be most happy to enter upon it.

R. But is not the word death, or deadness, an undesirable one to use?

W. To me it seems of all terms the best. But the

word is unimportant; if you object to it, dismiss it from your thoughts, and take up the question without it. This I say: Man feels that which is apart from him to be inert, not because it is as he feels it, but because of his own condition; if his feeling were true, he would feel himself in presence only of existence that is spiritual; it is through a want in him that his feeling is caused to be untrue. This is my position. Let the question of deadness be put aside. If I have erred in using the term, the issue raised and the arguments have just the same force and value, the same claim to be weighed and answered.

R. I think you have a right to take this ground. Perhaps I might as reasonably call on a geometrician to explain space before I would receive his demonstrations, as insist on your explaining why and how there is a want of life in man, before I admit the fact. The question is one of evidence, not of explanations.

W. I seek to place it on this ground; but it is not necessary to avoid the other, if the distinction between the proof and the explanation of a fact is borne in mind. I will tell you, briefly, what appears to me a possible view. May not the death of man be the loss of a lower, for the bestowment of a higher existence? may it not be a necessary condition of his life, because of the very nature of life as involving Love and Sacrifice? May not this deadness be a result of an act of sacrifice; itself part of the universal life?—death relatively to man, but absolutely life? Is not all life that we know based on death, and springing out of it? Is not every life purchased at the cost of other life? Except a grain of wheat fall into the ground and die, does it not abide alone? Has not science, also, taught us this? Life for life, it is the law of Nature. Why is not man's death a mode in which this law of love appears? But do not let these speculations influence you; my argument is the same, whether they be true or false.

R. There may be indications of a reasonable treatment of these subjects, and whatever thought is true must bring us into the presence of more and greater unsolved problems than that of which it gives the solution. If your view left none such, it were self condemned. But let me ask one more question. How is it that we see in geology so long a course of *physical* existence before man existed at all?

W. Do you really feel that to be a question?

R. I am half ashamed to ask it. You would say, that is the phenomenon. It is the spiritual that is physical to us.

W. In the same way we see how it is that we think and feel by physical brains and nerves. I base the proof of my position on its practical success in solving problems. This is ever the ultimate proof. Let a proposed view be put to the test, and see how it will work.—Surely it is the very *reductio ad absurdum* of the hypothesis of matter, that we must, on that view, attribute our consciousness to material changes in the brain as its true cause, and conceive that there is some place, or point, at which a mechanical impulse, or chemical process, becomes a sensation or a thought. But on the view that the physical is the mode in which the spiritual is perceived by man, this strange fact assumes an entirely new aspect. It ought to be; it might be foretold. For if the true cause of our consciousness is spiritual action on us, how should it appear to us? Evidently as physical action. Brain and nerves, or something equivalent, ought to be the 'phenomenal' instruments of consciousness, if this view be right.

R. I must think farther about this.

W. Let it pass. But is it not evident that man must, naturally, at first take a view of all things which makes himself the centre, and must he not afterwards attain a view which includes a recognition of his own position as subordinate? is it not necessary that he should first be confident in his own impressions, and afterwards learn to correct them by a consideration of himself? I seek only to take this step. It must, at some time, be taken.

R. I see that the latter mode of thinking is the more humble.

W. And is it not also the more rational? Is it possible that we can be right in continuing to set up ourselves as the standard of existence? No other course is possible at first, but the delusion exists only that it may be escaped from.

R. That which it has been right for men to do, formerly, it may be right for them to cease doing now?

W. Does it cast blame upon the past to say that it has prepared for a better future? You see what we do in thus altering our view. All phenomena remain the same, but we transfer our thought of *existence* from them, to something that is more and worthier than they. They are as they were before to our impressions and our use, but their relation to our thought is altered. We think of them not as the cause of man's experience, but as being present to his consciousness through the operation of a true cause, more real, more certain, more adapted to produce the known effects, than they.

R. This also I understand. Might you not

express your conception thus:—Physical things are to our touch but are not to our thought, as *appearances* are to our sight, but are not to our touch? Our thought should be of one thing, while our sensuous impression is of another. That which is to our sense is less than that which IS, and considered in and by itself, therefore, must be unreal.

W. So we constantly associate all our consciousness with the operation of existence above that which our impressions represent. Our thought and regard are ever directed to the spiritual, which alone we recognise as Cause. We live now in the spiritual world, and find it perfectly simple that the physical is but the phenomenon, and not the fact. The difficulty would be to think otherwise. To do so we should have to ignore the convictions that are most certain to us. To forget that we are in a world that is spiritual, surrounded by BEING that is Holy, we must put away the feeling of our own defectiveness, the assurance of God's infinitude, the consciousness that there is ACTION around us; we must silence the reason, stifle the conscience, crush the heart, enslave ourselves to sense against our better knowledge.

R. We act by appearances as existing to sight; may we not act by physical things as existing to touch, remembering, in each case, that that which exists is different from that of which we have the impression? And as we rationally explain our impressions of sight by our relation to that which is to our touch, so may we not seek a rational explanation of our impressions of touch in our relation to that true existence which we know must BE? If we find it so simple that we consciously *see* that which does not exist, may we not find it equally simple that we consciously *touch* that which does not exist?

W. Here you approach, again, the true secret of the difficulty that is felt in giving up the existence of the physical; our consciousness of action in it, and upon it. This it is that makes it real to us. Merely passive impressions, such as those of sight, we have no difficulty in understanding to have no existence corresponding to them. But if our active impressions, or those of touch, have truly no existence corresponding to them, then we must recognise defect in ourselves; we must regard our *self* in a different light. This is the very point of the argument. If physical things are demonstrated not to have true existence, then an unsuspected defect is demonstrated in man.

R. So your argument has been, to prove that that which is inert cannot be that which exists.

W. Yes. Keep that question fairly before you. If that which EXISTS cannot be inert, then it is defect in man which makes him feel himself in a physical world; he is not truly so. And all the practical inferences which follow claim a practical regard.

R. But there is one point more. In our action on the physical, all men alike perceive the results of that which each individual does. If I move anything, it is moved to all men's perceptions, not only to my own.

W. You do right to make this remark. The individual action has relation to man's universal conscious perception. It is not merely an individual, but an universal relation, that is involved. Deeper bonds unite men to their fellows, than upon our natural ideas we should have suspected.

R. Is not this another question which needs future investigation? Can you explain how this community of perception takes place, according to your view?

W. Only theoretically. And I especially wish to avoid weakening my argument by having recourse to theoretical explanations. I prefer to leave that question, altogether, as a matter for inquiry. I need not again remind you that unexplained circumstances do not weaken the force of a sound argument. One chief advantage of a truer mode of thinking is that it opens new channels to our thought. Nor need I point out to you that there are easy ways of reconciling this particular circumstance with that which I affirm, if we demanded plausibility alone.

R. Rather than give an explanation unsupported by proof, you prefer that it should stand as a circumstance not accounted for?

W. I do. The more unexplained facts we clearly apprehend, the more hopeful is our prospect of increasing knowledge. But think, if our entire experience is not such as it should be, if the inertness is due to man. Not one or two things, but all that we are conscious of, or perceive. Must there not be an unvarying phenomenon, under cause and effect, felt by us as real: must not this impress us with the feeling of force, correspondent to the inertness within; and could 'force' be other than such as it is in this phenomenon, which we call the physical? Does it not obey 'necessary' laws, laws conforming to the reasonable faculties in man? Must not force, indeed, be conformed to, and determined in its operation by, the resistance it implies? And is not this uniformity, of passive force controlled by force,—force as it were self-controlled—the very mode under which holiness should appear to a being to whose apprehension the action is wanting? Does it not speak to us, trumpet-tongued, of the control of passion in ourselves?

R. That is a striking fact. In nature no force, no passion, is uncontrolled.

W. It is an absolute rightness greets us there. Therefore we love it so, and trust it: the Manhood in us claims brotherhood with the Life around. Our uncontrolled enslaving passions, only, separate us from nature. And how well, and naturally we understand that a holy action, an act of Love and Rightness, is the sole cause of all that is. No accident to baulk, no passive law to crush, no deadness abhorrent to our souls, mocks us or constrains. One cause for all, alike for all. The hairs of our head are numbered, nor falls a sparrow to the ground without our Father. All is God's act and deed: weighted with the infinite Necessity which is His sole prerogative; constrained, but by His love alone; inevitable, but because He is Holy.

R. Should we say of each thing that affects us, each operation of which we are conscious on ourselves: This God does; His act is the cause of my feeling thus? and if we ask, why it is, answer ourselves: 'it must be by His nature'? and in tracing physical necessities, remember that we are tracing the evidence of his unchangingness?

W. Long ago was the question asked: Shall not the judge of all the earth do right? Science has answered it; He does.

R. That is not enough.

W. It is not. HE is not only holy. If righteousness looks down from heaven, truth springs up from the earth. Righteousness and peace have kissed each other. God gives life to man, His life for man. He has shown us what He does, and why. So we can rest and trust in Him. The reason of all things is that man must be redeemed. If in all our sorrows, all our joys, we could but think of that!

R. 'Tis time there came some change in our present thoughts. The world is tired of its endless round. Who is content?

W. I do not know. There are many who try to make themselves content, who think it a religious duty. But who will fairly look upon the world and say: I *am* content?

R. I would not be the man. Unless, indeed, it is true that God is redeeming man, and that all this history is the destroying of the death within him. If I could believe that I should be happy.

W. You would be. You could not help it. The power of an overwhelming joy would carry you along, compelling you to throw all your heart and soul into God's work. It would save you to believe; to believe in Christ, THE REDEEMER OF THE WORLD.

THE END.